Surveys and Tutorials in the Applied Mathematical Sciences

Volume 17

Series Editors

Anthony Bloch, Department of Mathematics, University of Michigan, Ann Arbor, USA

Charles L. Epstein, Department of Mathematics, University of Pennsylvania, Philadelphia, USA

Alain Goriely, Department of Mathematics, University of Oxford, Oxford, UK

L. Greengard, Department of Mathematics, Courant Institute of Mathematical Sciences, New York University, New York, USA

Advisory Editors

Michael Brenner, School of Engineering, Harvard University, Cambridge, USA

Gábor Csányi, Engineering Laboratory, University of Cambridge, Cambridge, UK

Lakshminarayanan Mahadevan, School of Engineering, Harvard University, Cambridge, USA

Clarence Rowley, MAE Department, Princeton University, Princeton, USA

Amit Singer, Fine Hall, Princeton University, Princeton, USA

Jonathan D. Victor, Weill Cornell Medical College, New York, USA

Rachel Ward, Department of Mathematics, University of Texas at Austin, Austin, USA

Featuring short books of approximately 80-200pp, Surveys and Tutorials in the Applied Mathematical Sciences (STAMS) focuses on emerging topics, with an emphasis on emerging mathematical and computational techniques that are proving relevant in the physical, biological sciences and social sciences. STAMS also includes expository texts describing innovative applications or recent developments in more classical mathematical and computational methods.

This series is aimed at graduate students and researchers across the mathematical sciences. Contributions are intended to be accessible to a broad audience, featuring clear exposition, a lively tutorial style, and pointers to the literature for further study. In some cases a volume can serve as a preliminary version of a fuller and more comprehensive book.

Jonathan Swinton

Mathematical Phyllotaxis

Springer

Jonathan Swinton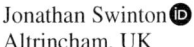
Altrincham, UK

ISSN 2199-4765   ISSN 2199-4773 (electronic)
Surveys and Tutorials in the Applied Mathematical Sciences
ISBN 978-3-031-94012-5   ISBN 978-3-031-94013-2 (eBook)
https://doi.org/10.1007/978-3-031-94013-2

Mathematics Subject Classification: 92C15, 92C80, 92C42, 37G10, 11A05, 92-10

© The Editor(s) (if applicable) and The Author(s), under exclusive license to Springer Nature Switzerland AG 2025

This work is subject to copyright. All rights are solely and exclusively licensed by the Publisher, whether the whole or part of the material is concerned, specifically the rights of translation, reprinting, reuse of illustrations, recitation, broadcasting, reproduction on microfilms or in any other physical way, and transmission or information storage and retrieval, electronic adaptation, computer software, or by similar or dissimilar methodology now known or hereafter developed.
The use of general descriptive names, registered names, trademarks, service marks, etc. in this publication does not imply, even in the absence of a specific statement, that such names are exempt from the relevant protective laws and regulations and therefore free for general use.
The publisher, the authors and the editors are safe to assume that the advice and information in this book are believed to be true and accurate at the date of publication. Neither the publisher nor the authors or the editors give a warranty, expressed or implied, with respect to the material contained herein or for any errors or omissions that may have been made. The publisher remains neutral with regard to jurisdictional claims in published maps and institutional affiliations.

This Springer imprint is published by the registered company Springer Nature Switzerland AG
The registered company address is: Gewerbestrasse 11, 6330 Cham, Switzerland

If disposing of this product, please recycle the paper.

For Adam

Preface

Over the past hundred years or more there has been a confusing variety of attempts to explain in mathematical language why Fibonacci numbers appear in plant spirals, but most modern accounts have now converged on a framework which might be called a Standard Picture. This book explains that mathematical framework and explores the extent to which it can explain Fibonacci phyllotaxis in the light of modern molecular biology. This Standard Picture relies on a biological model of how plants decide where to place individual new organs over the course of development, a model which when iterated tend to produce patterns we can model mathematically as lattices. Straight lines in these lattices are *parastichies*, with an associated parastichy number counting how many of the parastichies make up the lattice, and there is a natural way to model what we do when we visually identify spirals as obvious that selects two of these parastichy numbers as the *principal parastichy pair*. This enables us to classify many model patterns by pairs of integers. Then we change a parameter of the model slowly, corresponding to some slow change—almost always geometry—in the underlying developmental process. This gives us a bifurcation tree for principal parastichy pairs as that parameter varies. The culmination of the Standard Picture is that, under quite general assumptions, Fibonacci-like structures form stable branches of solutions.

It has been my goal in this book to equip readers to critique and participate in the applications of mathematical phyllotaxis. I aim to give you an understanding of enough of the underlying biology to motivate this Standard Picture, to analyse it mathematically, and to evaluate it scientifically. Actually the main goal was to learn about the subject myself, but in writing the text I have imagined the needs of a recently graduated student of mathematics embarking on research in an interdisciplinary laboratory. (Since I don't fit any of these criteria, I would be particularly grateful for critique from those who do.)

This is a textbook in which I have re-shaped and sometimes re-derived a lot of other people's work into a story I find coherent and worth sharing. Douady et al.'s recent book [36] provides a lively and accessible history of mathematical phyllotaxis showing how the subject has attracted a range of remarkable thinkers. I have not felt obliged to provide a source for every statement or idea, but notes to each Chapter situate it in the literature for further reading. One exception is that I have highlighted the contributions made by Alan Turing (and indeed his fiancé), reflecting the route by which I myself became fascinated by the subject.

The first half of the book is an account of the mathematical theory of two dimensional lattices. While lattice theory in higher dimensions has been an important area of mathematical research, the much simpler results we need in two dimensions are well-established, and I think intuitive. However my account is regrettably dense with Theorems and proofs. This was not my original aspiration, but evolved as I tried to make sense of multiple overlapping concepts, and occasional mathematical untruths, in the phyllotaxis literature. So there is in places a density of mathematical argument of the pedantic kind I learned to loathe when a graduate student many decades ago. To my surprise, in the end I rather enjoyed writing these parts, but for those whose taste does not run this way they are largely skippable. I have put less technical summary sections at the ends of these chapters.

Those whose taste *does* run pure-mathematically might find the exposition painful to read in a different way. Pure mathematicians will recognise entirely conventional ideas about co-primality and winding-numbers, cylindrical lattices as the quotient of planar lattices by a periodicity group, and hyperbolic geometry. If you are one of these pre-privileged readers I urge you to persist with my approach even though it does not always follow the classic presentations. Phyllotaxis, I believe, can offer an under-appreciated motivating example for teaching, because it traces a route from this pure mathematics to contemporary biological research issues.

The second half of the book connects the mathematical principles of the first half to biological observation by reviewing a range of mathematical models. This part reviews some basic plant embryology and molecular biology and introduces the apical meristem as the site of developmental and positional commitment. I will discuss older ideas, like Hofmeister's hypothesis that primordia are generated as soon as they are beyond a fixed distance from all other primordia, together with the modern molecular understanding of auxin-based mechanisms of primordia initiation. Then we can begin the applied mathematical work of evaluating how well our lattice models perform, and what they do and don't explain. In particular, we will see the mathematical and biological motivations for generalising lattice models to stacked coin models. Finally we'll review some of the outstanding mathematical questions, both pure and applied, and how they can inform the wider and continuing scientific debate on the generation of form in biology.

Altrincham/Cambridge, UK   Jonathan Swinton
March 2025

Acknowledgements In 2012, Mark and Ang Davis asked me to give a talk in the pub for Bollington Science Festival which started me wondering why it wasn't possible to explain why plants had Fibonacci numbers, even though I'd believed 20 years earlier that this was a mathematically solved problem. Around the same time, Erinma Ochu of the MOSI Turing's Sunflowers project was amazed not only that no-one knew how to breed Lucas sunflowers, but that I was surprised by the question. Julia Gog made some usefully mathematical remarks on a very early draft of the lattice theory chapters, and Paul Glendinning long ago gave some advice on how to phrase my challenge to the Fundamental Theorem of Phyllotaxis. Christopher Golé told me about his work with Stéphane Douady on the stability of rhombic lattices with three disks. Stéphane, together with Annemiek Cornelissen, made me welcome on a visit to the CNRS Laboratoire Matière et Systèmes Complexes. Tamsin Spelman and several of her colleagues at the Sainsbury Laboratory in Cambridge made helpful comments, and Hugo Tavares and Katie Abley's neighbourly support was particularly nourishing. Philip Maini graciously hosted me in the Mathematical Biology seminar at the University of Oxford and Andrew Krause, Mark Muldoon and Ronjoy Adhikari listened patiently and constructively to my pitches for this material. Phil Ramsden currently holds the record for the fastest finding of a mistake. Adam Swinton has been a patient and critical reader of several drafts. Diana Gillooly and David Tranah, as editors at two major university Presses, offered supportive, engaged and productive publication advice over a rather extended period. But in the end it was Alain Goriely who on his own initiative invited me to join the Springer Texts series, and I thank him and the Editorial Board for support for this book. Christopher Golé, Stéphane Douady and Richard Schwartz gave constructive advice as reviewers for Springer.

Thinking about this subject allowed me to rediscover a delight in mathematics. But love of mathematics alone would never have got this book finished over its many years in the writing. For rediscovering the love of almost everything else, including getting things done, I thank in particular Emma Anderson. But it was the questions of *all* these people and more, by email, in the pub, the museum and the seminar room, that shaped this attempt at answers. Thanks to you all.

Competing Interests The author has no competing interests to declare that are relevant to the content of this manuscript.

Contents

Part I Introduction

1 Motivation and Outline 3
   1.1 Fibonacci Phyllotaxis 3
   1.2 Joan Clarke's Daisy 3
   1.3 The Sunflower 4
   1.4 Fibonacci Structure Across the Plant Kingdom 6
   1.5 The Pineapple 7
   1.6 The Fir Cone 7
   1.7 Summary 10

2 Fibonacci Numbers, the Golden Ratio, and Co-prime Integers 11
   2.1 Fibonacci Sequences 11
      2.1.1 The Golden Ratio 11
      2.1.2 The Golden Angle 12
   2.2 Co-prime Integers 13

3 Co-primality, Continued Fractions, and Möbius Maps 15
   3.1 Bézout Relations 15
   3.2 The Winding-Number Pair 16
   3.3 Euclid's Algorithm 16
      3.3.1 Matrix Form of the Euclidean Algorithm 17
   3.4 Farey Trees and Winding-Number Pairs 19
      3.4.1 Fibonacci Pairs 21
   3.5 Continued Fractions 21
   3.6 Continued Fractions and Möbius Maps 22
      3.6.1 Möbius Maps 23
   3.7 Summary 25

Part II Mathematical Theory

4 The Geometry of Cylindrical Lattices 29
   4.1   Connections with Botanical Terminology 29
   4.2   Cylindrical and Plane Lattices 31
        4.2.1   Labelling Cylindrical Vectors 33
   4.3   Parastichies 34
        4.3.1   Visible Points and Parastichies 35
   4.4   Parastichy Pairs 36
        4.4.1   Opposed Parastichy Pairs 37
   4.5   Generating Pairs 37
        4.5.1   Generating Pairs as Basis Vectors 39
   4.6   Estimating the Divergence for a Generating Parastichy Pair 41
   4.7   Opposed Intervals 42
   4.8   The Fundamental Theorem of Phyllotaxis 42
   4.9   Principal Parastichies 45
   4.10 Opposed and Non-opposed Lattices 46
   4.11 Spiral Lattices 47
   4.12 Turing-Euclid Reduction of Generating Pairs 48
   4.13 Orthostichies 50
   4.14 Touching-Circle and Hexagonal Lattices 51
   4.15 Multijugate Lattices 53
   4.16 Dropping the Hats 53
   4.17 Relation to *Phyllotaxis: A Systemic Study* 54
   4.18 Summary 56

5 Classifying Cylindrical Lattices 57
   5.1   Components of the Van Iterson Diagram: The m=n Branch 57
   5.2   Triple-Points 60
        5.2.1   The m=n Branch in Lattice Space 61
   5.3   Nearly Hexagonal Lattices: Unfolding the Triple-Point
         Bifurcation 61
   5.4   Lattice Renormalisation 65
   5.5   Classification of Van Iterson Space by Euclidean Reduction 69
   5.6   Packing Efficiencies 69
   5.7   The Structure of Multijugate Lattice Space 72
   5.8   Notes to this Chapter 72
   5.9   Summary 74

6 Transformed Lattices 77
   6.1   Smoothly Changing Lattices 77
   6.2   Lattices on Disks 79
        6.2.1   Exponential Scaling 80
        6.2.2   Quadratic Scaling 80
        6.2.3   Biological Choices 80
   6.3   Other Arenas 82
   6.4   Notes to this Chapter 84

Part III Mathematical Modelling

7 Developmental Biology of the Plant Stem 87
  7.1 Stem Extension and Thickening 87
  7.2 Developmental Commitment of Primordia 89
  7.3 Cellular Architecture of the Shoot-Apical Meristem 91
  7.4 Molecular Phyllotaxis 91
  7.5 Mechanical Stress 95
  7.6 Vascular Bundles and Leaf Traces 95
  7.7 The Relationship to the Standard Picture 95
  7.8 Development of the Capitulum of the Sunflower 96

8 Statistical Phyllotaxis 101
  8.1 Early Idealism 101
  8.2 Observational Phyllotaxis 102
    8.2.1 Statistical Phyllotaxy and Rare Parastichy Pairs 103
    8.2.2 Statistical Phyllotaxis of Fir Cones 105
  8.3 Non-lattice Patterns 107
    8.3.1 Loss of Symmetry 107
    8.3.2 Dislocations 107
    8.3.3 Rising and Falling Phyllotaxis 109
  8.4 Beyond Parastichy Counts 110
    8.4.1 Bract and Ray Floret Counts 110
    8.4.2 Column Patterns 111
  8.5 Summary 112

9 Placement Models 113
  9.1 The Douady-Couder Ferromagnetic Model 113
  9.2 Energy Based Models 116
  9.3 Auxin Flow Models: Partial Differential Equation
       and Cell-Based Models 117
    9.3.1 An Early Reaction-Diffusion Approach 117
    9.3.2 Cell-Explicit Auxin Flow Models 118
  9.4 Mechanical Tension Models 119
  9.5 A Divergence: Protractor Models 119
  9.6 L-System Models 120
  9.7 Summary 121

10 Stacked-Disk Models 123
  10.1 Motivation 123
  10.2 Stacked-Disk Models 124
    10.2.1 Stacked-Disk Models Cannot Generate
            Non-Opposed Lattices 126
    10.2.2 Parastichy Numbers in Stacked-Disk Models 126
    10.2.3 Stacked-Disk Models Naturally Generate
            Fibonacci Transitions 128

10.3	Outstanding Questions	129
10.4	Coherent Structures in Stacked-Disk Models	130
	10.4.1	Cylinder Tilings and Rhombic Tilings	131
10.5	Columnar Models	133
10.6	Finite-Time Dynamics and Sunflower Data	135
	10.6.1	Cylinder to Capitulum Mappings	136
	10.6.2	Falling Phyllotaxis and Observed Asymmetry on the Capitulum	137
10.7	Notes to this Chapter	138

Part IV Conclusions

11 Some Future Directions 143
   11.1 Mathematical Issues 143
      11.1.1 The Van Iterson Diagram 143
      11.1.2 Cylindrical Tilings 143
      11.1.3 Regular Patterns as Attractors 144
   11.2 Modelling Challenges 144
      11.2.1 The Phyllotactic Jump 144
      11.2.2 Macroscopic Tests of the Standard Picture 145
      11.2.3 Universality Across Phyla 146
   11.3 Conclusion 147

12 Answers to Exercises 149

References 161

Index 169

Part I
Introduction

Chapter 1
Motivation and Outline

Abstract This chapter introduces the observation of Fibonacci phyllotaxis: that the arrangement of leaves, seeds, or scales in some plants can come in spirals, and the count of these spirals is often a Fibonacci number. There are many accessible example of Fibonacci phyllotaxis: perhaps the most robustly repeatable observation for city-dwellers is to buy a pineapple. However for most species, including the pineapple, large-scale scientific data is surprisingly sparse on the exact prevalence of different kinds of patterns, and especially on the ways in which patterns fail to be Fibonacci. An argument of this book is that it is when patterns fail to be Fibonacci that they are most informative for model testing, and we will return to this in the second half of the book.

1.1 Fibonacci Phyllotaxis

The Fibonacci series is

$$0, 1, 1, 2, 3, 5, 8, 13, 21, 34, 55, 89, 144, \ldots$$

where each number after the second is the sum of the previous two. Phyllotaxis means the arrangement of the organs of plants, like leaves and seeds, and Fibonacci phyllotaxis is a reference to the appearance of Fibonacci numbers in different kinds of counts of these organs.

1.2 Joan Clarke's Daisy

For a few months in the summer of 1941, a young British couple were engaged to be married. Both were talented and overworked cryptographers working at Bletchley Park, the wartime codebreaking station, and Joan Clarke and Alan Turing would spend their off-duty hours together, playing chess, going for bicycle rides and visiting each other's parents. They broke the engagement off after a disastrous wet weekend in Wales, but when in the 1970s Turing's biographer Andrew Hodges interviewed Clarke, what she remembered distinctly about her long summer with her fiancé

Fig. 1.1 Thirteen green bracts on the underside of a daisy *Bellis perennis*, picked in 2019 from the Bletchley Park lawn where Joan Clarke taught Alan Turing about phyllotaxis in the summer of 1941

was lying on the lawn in front of Bletchley Park looking at the daisies. Turing and Clarke already knew about and shared an enthusiasm for the appearance of Fibonacci numbers in plants and though both were first-class mathematicians, it was Clarke who had to teach Turing about the botanical classifications of plant structure. So perhaps she pulled up one of the summer daisies and explained that the distinct green 'petals' on the underside of the flower should be called bracts, but either way, when they counted the bracts they will most likely have found thirteen of them, as I too did in the summer of 2019 (Fig. 1.1).

Clarke went on to be a significant figure in post-war British cryptography, but it is Turing's legacy which is now famous both inside and outside mathematics. Undergraduates studying mathematical biology commonly encounter his Turing instability as a mechanism for pattern formation, but what is rarely taught is that one of the unpublished motivations for his published theory was to explain the appearance of Fibonacci numbers. Indeed, at his death he left behind a manuscript scrap, *Outline of the Development of the Daisy*, and it was trying to interpret this scrap that first interested me in this absorbing problem. Turing died tragically before he fully worked out his theory, although thirty years later other mathematicians, not aware of his work, would take more or less the same direction to reach the mathematics which we are now able to outline in this book.

1.3 The Sunflower

As part of the Turing birth centenary celebrations in 2012, the Museum of Science and Industry in Manchester invited members of the public across the globe to grow sunflowers, *Helianthus annuus* and send them, or pictures of them, into the museum, as a citizen science experiment to determine the prevalence of Fibonacci numbers in the spirals of the sunflower head. Figure 1.2 shows one such submission, with different families of spirals visible to the eye. The three families of spirals highlighted in the Figure have 34, 55, or 89 members: these are the *parastichy counts*. Which family is most visually prominent is a subjective question, and one with a different answer at different places in the sunflower.

The mature seedhead retains evidence of the relative position on the embryonic plant of the cell lineages which were eventually committed to develop into a seed. The centre of the seedhead is newest in the sense that pattern on the outside of the

1.3 The Sunflower

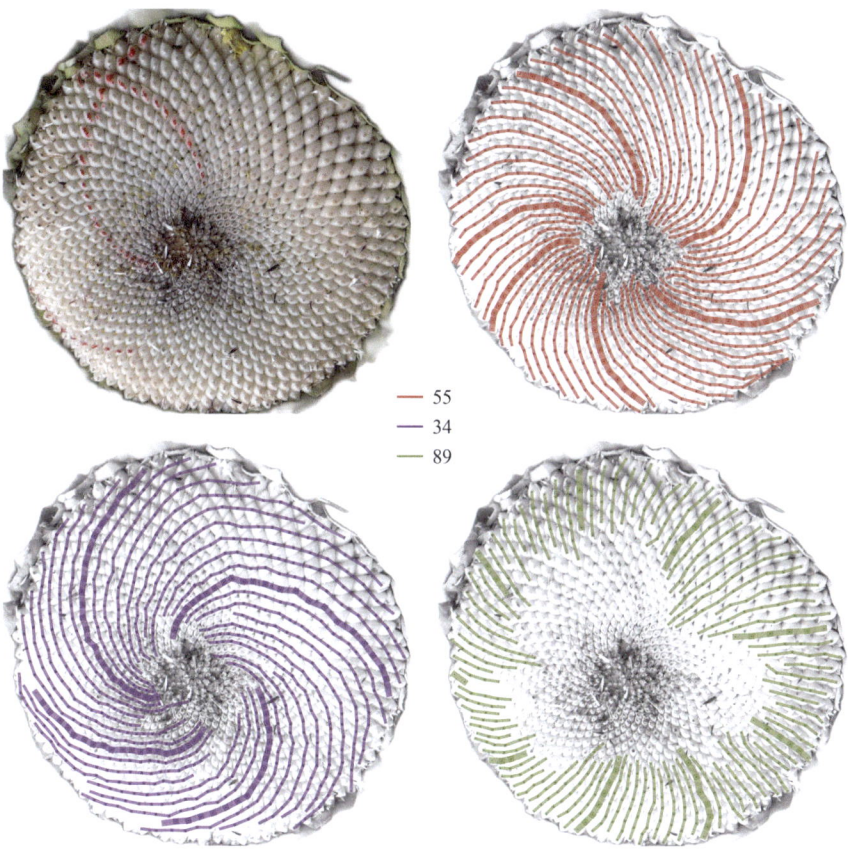

Fig. 1.2 Seeds packed onto the head of a sunflower. Three separate sets of spirals have been identified. For each set every tenth spiral is marked with a thick line. To me the most visually obvious spiral set in photograph is the red set of which there are 55, curving anticlockwise as they move outwards, the next the set of 34 clockwise blue spirals, but the set of 89 green spirals is also prominent towards the outside and especially in the upper right hand region. This sunflower head was labelled as sample 91 in [116], and is typical of the more ordered arrangements described there

sunflower is laid down before that towards the centre, and as we move into the centre, the seeds are the ones which have arisen from the most recent decisions to commit. In this Figure the most prominent pair on the outer rim is (55, 89) as we move inwards and forwards in developmental time this changes to the smaller pair (34, 55): this is called a *falling phyllotaxis*.

These parastichy counts on the sunflower, like 55 and 89, occasionally 144 and very occasionally 233, are the largest Fibonacci numbers which can reliably be observed in any plant forms. A common Just So explanation for these observations is that they provide some form of selection advantage for the plant in allowing an 'optimal' packing of the seedhead. It is clear from the prevalence of Fibonacci numbers in many species subject to natural rather than human selection that this is no artefact of human preference, but it is worth noting that these exceptionally large numbers are typically *not* seen in populations subject to natural selection, but in the giant forms of the sunflower, with seedheads 30 cm or more across, which have been under intensive breeding pressure by humans over the ten thousand years or so that we have been relying on the sunflower as a food source. The evidence of the archaeological record is that both the sunflower seed, and the capitulum in which it sits, have grown larger over the development of agriculture, and that this has happened more than once in different human cultures [20, 72]. It is likely that any selection pressure was for increased overall seed-mass, rather than any direct human desire for 'optimal packing'.

Daisies and sunflowers are close relatives within what modern taxonomy calls the *Asteracaea* family, and an older name for this family, the *Compositae*, reflects the common structure of most of them, with many individual flowers packed together on a flat composite seedhead or capitulum. Other members of the family, such as the dahlia, can also exhibit pairs of Fibonacci counts. However Fibonacci phyllotaxis is not at all a feature unique to this family.

1.4 Fibonacci Structure Across the Plant Kingdom

Among flowering plants, Fibonacci patterns are also relatively common in the kinds of cacti sold as houseplants in the UK: they can sometimes be seen in, for example the prickly pear. The spurges (*Euphorbiaceae*) are a genus of shrubs which are interesting in demonstrating rising phyllotaxis in a fairly clear way. One of the earliest systematic British studies of Fibonacci patterns was carried out by the Oxford botanist Arthur Harry Church, and self-published by him from 1904. Figure 1.3 shows the stem of a spurge growing outside his office in the Oxford Botanic Garden.

Fig. 1.3 Four successive sections of the stem of a *Euphorbia wulfenii* in the Oxford Botanic Garden, photographed by AH Church in 1904. From [24]

Church has removed all of the former leaves, and in effect numbered each of their positions rising up the stem. Thus when he writes a 3 on the first specimen, he is implying that there are two other branch positions, which we can't see behind the stem. The lines drawn by Church in Fig. 1.3 are meant to suggest that the branch positions are organised in spirals making up families with, in one case, 13 and 21 members. Church's book, though dated in places, remains the most valuable compendium of Fibonacci (and non-Fibonacci) structure in phyllotactic counts and is the basis for much of this chapter. For example he also records that pairs of Fibonacci counts like 5 and 8 can also be seen in the leaf structure of arums (*Arum*), teasels (*Dipsacus*), and crassulas (*Crassulacae*).

1.5 The Pineapple

A convenient example of Fibonacci patterning is often available to the supermarket consumer on the form of the pineapple.

1.1 Show this. It is *not enough* to believe that you can find them if you look: you have to look. In the absence of a local grocer, consult Fig. 1.4.

Despite the easily visible presence of Fibonacci spirals in pineapples, Church's systematic 1904 survey of phyllotaxis includes no pineapples or any other representative of their family, the bromeliads. This may have been because at the time whole pineapples were very rarely grown in Europe except as a luxury item, and most pineapple consumed was transferred to Western consumers as a chopped and tinned item from colonial plantations. Indeed, it was not until 1933 that the observation of Fibonacci counts in pineapples made it into the scientific literature, when Linford, working on a commercial plantation, reported the occurrence of Fibonacci counts in a discussion of how to efficiently count the number of eyes [75]. The advent of whole fruit transport in the 1950s, enabled mathematical popularisation to mention the pineapple, notably in the work of Hal Coxeter, and with a suitable meaning for the word 'natural', the pineapple has come to rival the sunflower as an emblem of 'mathematics in the natural world'.

1.6 The Fir Cone

Fibonacci structure is not restricted to the flowering plants and can also be found in the conifers. Figure 1.5 shows a fir cone, one of six thousand collected from a black pine tree on the slopes above Lake Zurich in a remarkable experiment by Dr. Veronika Fierz. As in the Museum of Science and Industry experiment, the main scientific motivation was to explore deviations from strict Fibonacci structure, but as with sunflowers the data also shows how prevalent Fibonacci structure was: all

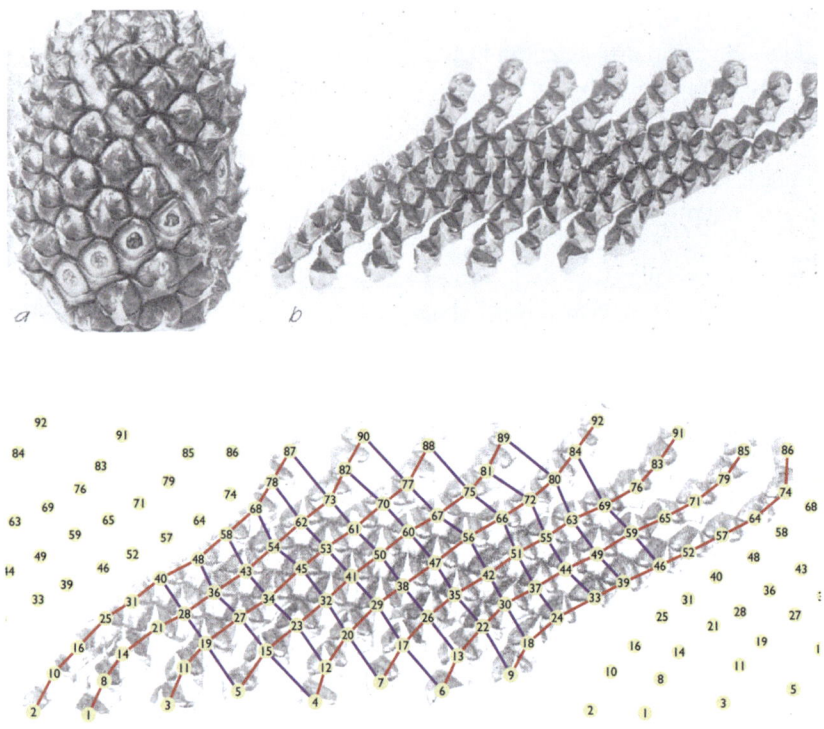

Fig. 1.4 Top: **a** whole pineapple and **b** flattened pineapple skin from [75]; the black lines are natural markings on the scales. Bottom: as **b**, but with the centre of each scale numbered in order of height in the photograph, and physically adjacent scales joined by lines

but a few hundred had eight spirals of the seed-protecting scales in one direction and thirteen in the other.

1.2 Use Cecilia Braun's[1] drawings (Fig. 1.6) of pine cones with the scales numbered to count the spirals they display.

Fibonacci counts, typically but not always 3 s, 5 s and 8 s, can be found in the fruit cones of other conifers, including spruces (*Picea*), larches (*Larix*), cypresses (*Cupressus*) [39] and the monkey-puzzle tree (*Araucaria*) [24].

---

[1] Cecila Suzette Agassiz (1809–1848) had been a pupil of the German painter Maria Ellenrieder and would herself 'have become an artist of repute had she devoted her life to the fine arts' [77]. She provided all of the drawings in her brother Alexander's 1831 book on pine cone phyllotaxis [17] although only his name appeared on the title page. At the same time, she cared for Alexander's university friend Louis Agassiz during a long bout of typhoid fever. In 1833 she married Agassiz and until the arrival of 'family duties' provided illustrations for Agassiz's many works [77].

1.6 The Fir Cone

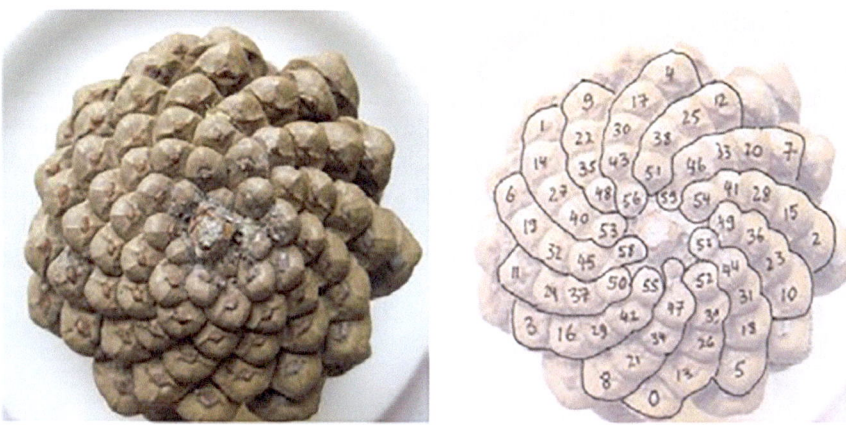

Fig. 1.5 One of over 6000 pine cones of a single *Pinus negra* studied by Fierz, of which 97% had, like this specimen, eight spirals visible in in direction and thirteen in the other. The cone was attached to the tree through the central region, with the growing tip furthest away from the camera, so that the scales are numbered by youth, with scale 0 developing after the others. The thirteen spirals are outlined directly; that there *are* eight in the other direction can be confirmed by picking out eg the sequence 2, 10, 18, 26, 34, ... and seeing we can do this eight times. From [39]. ©CC-BY 3.0 Fierz, 2015

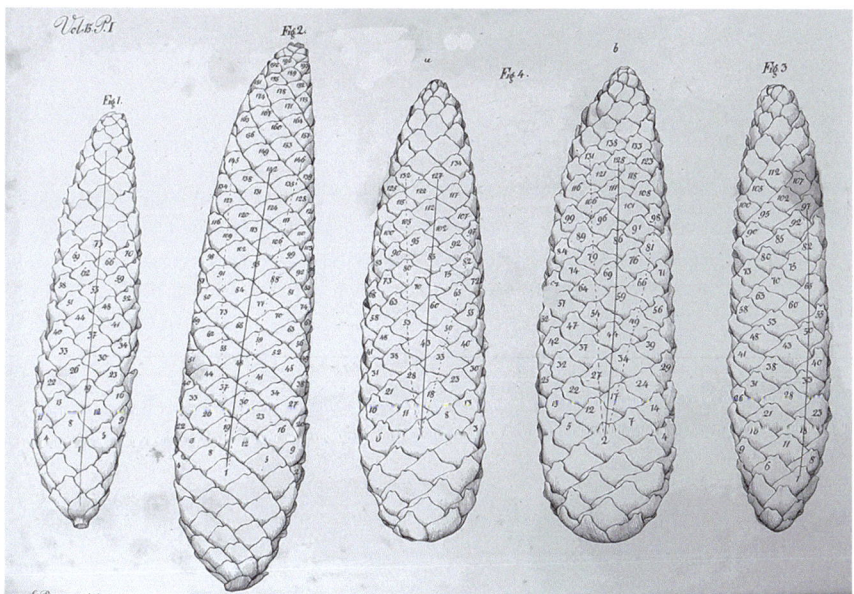

Fig. 1.6 Spruce *Picea* cones from from Plate XXVII of Braun's 1831 book [17]; Fig. 4(b) is the back view of Fig. 4(a)

1.7 Summary

There are many accessible example of Fibonacci phyllotaxis: perhaps the most robustly repeatable observation for city-dwellers is to buy a pineapple. However for most species, including the pineapple, large-scale scientific data is surprisingly sparse on the exact prevalence of different kinds of patterns, and especially on the ways in which patterns fail to be Fibonacci. An argument of this book is that it is when patterns fail to be Fibonacci that they are most informative for model testing, and we will return to this in the second half of the book.

Chapter 2
Fibonacci Numbers, the Golden Ratio, and Co-prime Integers

Abstract The Fibonacci sequence F_n has as its first two members $F_0 = 0$, $F_1 = 1$ and every subsequent member is the sum of the previous two: $F_{n+2} = F_{n+1} + F_n$. Although there is a substantial literature on Fibonacci and related sequences we really only need this simple sum property and some elementary related facts collected here.

2.1 Fibonacci Sequences

The Fibonacci sequence F_n has as its first two members $F_0 = 0$, $F_1 = 1$ and every subsequent member is the sum of the previous two: $F_{n+2} = F_{n+1} + F_n$. A related sequence is the Lucas[1] sequence $(1, 3, 4, 7, 11, \ldots)$ with the same rule but different initial conditions. Both Fibonacci and Lucas numbers are special cases of the generalised Fibonacci numbers with starting pair $F_1^k = 1$, $F_2^k = k$. We will also encounter the sequence created by doubling each Fibonacci term, for which terminology varies but I'll call a double-Fibonacci sequence.

We might note that from Table 2.1 that the Fibonacci, double Fibonacci and Lucas sequences together include all of the first eleven integers except 9, so there is little remarkable about the observation that a particular system exhibits a structure including a low member of one of the sequences [27]. Finding *pairs* of adjacent members of the same series, or examples of numbers greater than ten, might be more significant.

2.1.1 The Golden Ratio

Starting the sequence with $F_0 = 0$, $F_1 = 1$, the general Fibonacci term is

$$F_n = \frac{\tau^n - (1-\tau)^n}{\sqrt{5}} \qquad (2.1)$$

---

[1] Named after Édouard Lucas, a French mathematician [9, 31], and so best pronounced *Luke–aah* by English speakers.

© The Author(s), under exclusive license to Springer Nature Switzerland AG 2025
J. Swinton, *Mathematical Phyllotaxis*, Surveys and Tutorials in the Applied Mathematical Sciences 17, https://doi.org/10.1007/978-3-031-94013-2_2

Table 2.1 Various sequences with Fibonacci structure

Fibonacci	1, 1, 2, 3, 5, 8, 13, 21, 34, 55, 89, 144, ...
Double Fibonacci	2, 2, 4, 6, 10, 16, 26, 42, 68, 110, ...
Lucas (F^3)	1, 3, 4, 7, 11, 18, 29, 47, 76, 123, 199 ...
F^4	1, 4, 5, 9, 14, 23, 37, 60, 97, 157 ...
F^5	1, 5, 6, 11, 17, 28, 45, 73, 118, 191 ...
...	
F^8	1, 8, 9, 17, 26, 43, 69, 112, 181 ...

where τ is the golden ratio satisfying

$$\tau^2 = \tau + 1 \tag{2.2}$$

$$\tau = \frac{1+\sqrt{5}}{2} \approx 1.618 \tag{2.3}$$

$$= \lim_{n\to\infty} \frac{F_{n+1}}{F_n}. \tag{2.4}$$

The sequences in Table 2.1, and indeed any sequence obeying the Fibonacci rule, all have τ as the limit of the ratio of terms. These results can be found by trying solutions of the form $F_n = k\tau^n$ and seeing that τ must be one of the two solutions of (2.2). This book only needs the facts in this chapter, but there is vast literature on Fibonacci and related sequences of which Vajda [125], say, is an accessible example.

2.1.2 The Golden Angle

The golden angle is

$$\Phi = \frac{2\pi}{\tau^2} \approx 137°$$

so

$$\frac{\Phi}{2\pi} \approx \frac{F_n}{F_{n+2}}.$$

As we'll see, the angular rotation between successive leaf structures is often very close to this angle for Fibonacci (but not, for example, Lucas) phyllotaxis.

2.2 Co-prime Integers

Two integers (m, n) are co-prime iff[2] their greatest common divisor is equal to 1, so that for example 4 and 11 are co-prime. There are some fiddly edge cases: $(0, 1)$, $(1, 1)$ and $(1, n)$ are co-prime for integer $n > 1$, but $(0, 0)$ and $(0, n)$ are not. These cases are consistent with the observation that every integer is a divisor of 0, but the only divisors of 1 are 1 and -1. S In the next chapter we will review the classic machinery for computing co-primality: the Bézout relationship and the Euclidean algorithm.

---

[2] The word 'iff' is not a typo. It means if and only if, and is also a shibboleth. If it is bewildering to you that 'iff' means logical equivalence and not just implication then you have not been inducted into the kind of mathematical exposition, roughly first-year undergraduate, we use until around Sect. 5.9, so you will want to skip to there. On the other hand you can read all of the rest of the book without knowing what a shibboleth is.

Chapter 3
Co-primality, Continued Fractions, and Möbius Maps

Abstract It would be wise to skip, or read very glancingly, this chapter on a first reading. The material on co-prime integers and Bézout relations is only needed for those who want to closely follow the proofs of Chap. 4. Möbius maps are the main language of the renormalisation approach to the van Iterson diagram discussed at the end of Chap. 5, but are not at all essential to understand the biological implications of the mathematics. The details of the matrix form of the Euclidean algorithm may only matter to those curious about the relationships between Fibonacci phyllotaxis and Escher prints.

3.1 Bézout Relations

The previous chapter defined co-primality for a pair of integers. An equivalent condition is the existence of a Bézout relation: the integer pair (u, v) satisfies the Bézout relation for the non-negative integer pair (m, n) iff

$$|nu - mv| = 1, \tag{3.1}$$

or equivalently as a matrix determinant:

$$\det \begin{pmatrix} n & v \\ m & u \end{pmatrix} = \pm 1. \tag{3.2}$$

So $(3, 4)$ are a Bézout pair for $(8, 11)$. If u and v exist then they are a certificate of co-primality: we will see below that (m, n) can satisfy $nu - mv = g$ iff $|g|$ is the greatest common factor of m and n. They are not unique because if (u, v) satisfies the Bézout relation then so does $(u + km, v + kn)$ for any integer k: $(11, 15)$ is also a Bézout pair for $(8, 11)$. We can introduce a range condition by picking a particular k which can be used to enforce $0 \leq v < n$, but there is still a further ambiguity because if $nu - mv = +1$ provides one solution, then $n(m - u) - m(n - v) = -1$ provides another: because $11 \times 3 - 4 \times 8 = +1$ we can find $11 \times 1 - 4 \times 3 = -1$.

3.2 The Winding-Number Pair

There are different ways to cope with the non-uniqueness of Bézout pairs. Here we pick out a particular pair, the *winding-number pair*. For co-prime integers (m, n), the winding-number pair is the unique pair (u, v) which satisfies the Bézout relation $|nu - mv| = 1$ and also

$$0 \le \frac{u + v}{m + n} \le \frac{1}{2}. \qquad (3.3)$$

There are a handful of special cases: in this book the winding-number pair for $(0, 1)$ is by definition $(1, 0)$, for $(1, 0)$ is $(0, 1)$, and for $(1, 1)$ is $(1, 0)$. As we will see below, half of the time the winding-number pair will yield $nu - mv = +1$ and half the time $nu - mv = -1$. We could alternatively have found a unique pair by insisting on a particular choice of sign $nu - mv = +1$, but in this way there is a clear connection to the signs of the Δs that will pepper Chap. 4.

3.1 Compute the winding-number pair for (m, n) for co-prime m, n less than 10.

Chapter 4 will also reveal the reason for the name 'winding-number pair'. For the small integers in plant spiral counts it is perfectly feasible to compute highest common factors and winding-number pairs by exhaustive search. But nevertheless the rest of this Chapter shows how an algorithmic approach sheds a light on the structure of co-prime pairs and their winding-numbers in a way that has been helpful in the past to mathematicians puzzling over Fibonacci structure.

3.3 Euclid's Algorithm

Euclid's algorithm finds the highest common divisor of two positive integers by repeatedly subtracting the smaller as many times as we can from the larger to find a new, smaller pair of integers, stopping when one of them is zero. It is simple to implement, and tracking the variables in the intermediate steps of the algorithm will allow us to understand connections with continued fractions and structure of the van Iterson diagram of Chap. 4. For example, consider calculating the highest common factor of 4 and 11:

$$11 - 2 \times 4 = 3. \qquad (3.4)$$
$$4 - 1 \times 3 = 1. \qquad (3.5)$$
$$3 - 3 \times 1 = 0. \qquad (3.6)$$

This works by successively answering the question of how many times 4 goes into 11 (i.e. 2), 3 into 4 (1), and 1 into 3 (1) and generates a particular sequence of what we will call $q_i = 2, 1, 3$ which shows the co-primality of 4 and 11. Because of the

3.3 Euclid's Algorithm

central role of the q_i they are coloured in red in what follows. It is possible to use this decomposition to compute the winding-number pair, which we can see if we rewrite the algorithm more formally. Given integers $n \geq m > 0$ we set $r_{-1} = n$, $r_0 = m$ and $i = 0$:

1. Set an integer $q_i = \lfloor r_{i-1}/r_i \rfloor$, to non-negatively minimise $r_{i-1} - qr_i$.
2. Set $r_{i+1} = r_{i-1} - q_i r_i$.
3. If $r_{i+1} = 0$, set $N = i$ and terminate.
4. Otherwise increment i, and repeat.

Theorem 3.1 *Euclid's algorithm terminates with the greatest common factor $GCF(m, n)$:*

$$r_N = GCF(m, n). \tag{3.7}$$

Proof The r_i are a strictly decreasing sequence of positive integers and so the algorithm always terminates. Suppose the GCF is k. Now k divides r_0, $k|r_0$, and the ith step of the iteration preserves the fact that $k|r_i$, and so in particular $k|r_N$. Now $r_N|r_{N-1}$ and the iteration step also shows that if $r_N|r_i$ and $r_N|r_{i-1}$ then $r_N|r_{i-2}$, and so following the r_is in reverse order we see r_N divides all of them and divides both m and n. So r_N is a common divisor and so $r_N \leq k$ but since $k|r_N$, $r_N = k$. □

3.3.1 Matrix Form of the Euclidean Algorithm

We can solve the Bézout relation by putting Euclid's algorithm in matrix form. For example the first reduction of (3.5) is

$$\begin{pmatrix} 11 \\ 4 \end{pmatrix} = \begin{pmatrix} 1 & 2 \\ 0 & 1 \end{pmatrix} \begin{pmatrix} 3 \\ 4 \end{pmatrix}$$

$$= \begin{pmatrix} 1 & 1 \\ 0 & 1 \end{pmatrix}^2 \begin{pmatrix} 0 & 1 \\ 1 & 0 \end{pmatrix} \begin{pmatrix} 4 \\ 3 \end{pmatrix}$$

$$= E^2 S \begin{pmatrix} 4 \\ 3 \end{pmatrix},$$

where we have defined matrices E corresponding to a 'Euclidean' reductions and S corresponding to 'swap' of the integer pair. Matrix algebra shows E^q has a q in the upper-right corner:

$$E^q = \begin{pmatrix} 1 & q \\ 0 & 1 \end{pmatrix} ; S = \begin{pmatrix} 0 & 1 \\ 1 & 0 \end{pmatrix} ; E^q \cdot S = \begin{pmatrix} q & 1 \\ 1 & 0 \end{pmatrix}.$$

Moreover $\det E = 1$ and $\det S = -1$ and so the matrix of any product of Es and Ss has determinant of modulus 1.

18   3 Co-primality, Continued Fractions, and Möbius Maps

After we learn the q sequence through the Euclidean algorithm we can write the result in matrix form as

$$\begin{pmatrix} 11 \\ 4 \end{pmatrix} = (E^2 S) \cdot (E^1 S) \cdot (E^3 S) \begin{pmatrix} 1 \\ 0 \end{pmatrix}.$$

If we apply the same series of transformations to the column vector $(0, 1)$ we will obtain a column vector of the form (v, u) so that

$$M_{213} = \begin{pmatrix} 11 & v \\ 4 & u \end{pmatrix} = (E^2 S) \cdot (E^1 S) \cdot (E^3 S) \begin{pmatrix} 1 & 0 \\ 0 & 1 \end{pmatrix}. \quad (3.8)$$

This shows there is a u and v which satisfy $|4v - 11u| = 1$ and give us a Bézout relation, and since $\det E = 1$ but $\det S = -1$, the sign of the determinant of M is determined by how many S matrices appear in the product. It also provides a way to calculate u and v:

$$M_{213} = \begin{pmatrix} 2 & 1 \\ 1 & 0 \end{pmatrix} \cdot \begin{pmatrix} 1 & 1 \\ 1 & 0 \end{pmatrix} \cdot \begin{pmatrix} 3 & 1 \\ 1 & 0 \end{pmatrix}$$

$$= \begin{pmatrix} 11 & 3 \\ 4 & 1 \end{pmatrix}.$$

For a general m and n, using the q_i values from the Euclidean algorithm gives us

$$M_{q_1 q_2 \ldots q_N} = \begin{pmatrix} n & v \\ m & u \end{pmatrix} = (E^{q_1} S) \cdot (E^{q_2} S) \cdots (E^{q_N} S) \begin{pmatrix} 1 & 0 \\ 0 & 1 \end{pmatrix}. \quad (3.9)$$

Because every pair of co-prime integers has a unique q_i sequence, every pair such pair will appear exactly once in the first column of a matrix in the tree generated by continuing Fig. 3.1 downwards. The second column is the corresponding Bézout pair. The same algorithm also works when m and n are not co-prime:

3.2 Show that if m and n have highest common factor g then u, v found in this way give $|nu - mv| = g$.

As we will see in the next section, this Bézout pair is in fact the winding-number pair.

3.3 Show that for $j > 1$ the q_i for the Fibonacci pair F_{j+1} and F_j are a sequence of 1 s of length j and the pair have a Bézout pair, but not a winding-number pair, F_j and F_{j-1}.

For those comfortable thinking of vector spaces, we might think of any pair of co-prime integers as a basis pair for the space of integers. Given a co-prime pair m and n, every integer can be expressed as a sum $n(ku) + m(-kv)$. From this perspective the Euclidean algorithm can be seen as a series of changes of these bases. The individual E^q matrices correspond to the change of base at each step, the product matrix is the

3.4 Farey Trees and Winding-Number Pairs

Thanks to a careful choice of stopping rule for the algorithm, or equivalently because of how we pruned the tree of Fig. 3.1 not to bifurcate from the first two nodes, the (u, v) which emerge from the Euclidean algorithm are not just a pair satisfying the

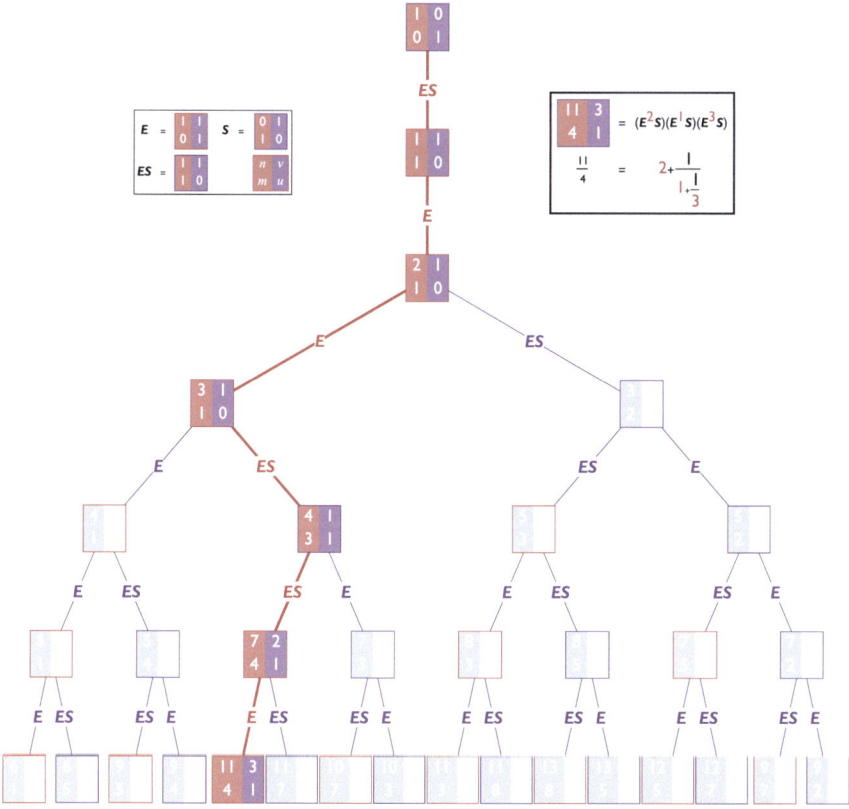

Fig. 3.1 Computing the highest common factor of 4 and 11 by successive matrix reductions upwards to the identity matrix. Conversely, the extended tree downwards from the identity matrix, branching by an E or an ES matrix-multiplication at each node after the second, will contain every pair of co-prime integers, in the first column of each matrix, with the corresponding winding-number pair in the second column. The q_is of the Euclidean reduction emerge as the number of Es between each S. The horizontal order of the E and ES branches below each matrices is chosen so that the child matrices are ordered by their Farey mediant as defined in Sect. 3.4. Chapter 5 will reflect on the close similarity between this diagram than the van Iterson classification

Bézout relation but they are exactly the winding-number pair. Each matrix in the tree of Fig. 3.1 is of the form

$$\begin{pmatrix} n & v \\ m & u \end{pmatrix}$$

with $nu - mv = \Delta$ and $|\Delta| = 1$. For each matrix below the root, we can form two rationals u/m and v/n; for $\Delta = 1$, $u/m > v/n$ while for $\Delta = -1$, $u/m < v/n$ but in either case they form the endpoints of a real interval we will call the Farey interval for the matrix. The Farey interval of the integers m, n, u, v is $[u/m, v/n]$ although the endpoints may not be in order. The Farey sum of these two rational endpoints, or the Farey mediant of the matrix, is $(u+v)/(m+n)$ which is always contained in the Farey interval.

3.4 Show this.

Farey intervals for the first few matrices in the tree are shown in Fig. 3.2. If a matrix has a Farey interval of $[u/m, v/n]$, then multiplication by E gives a new matrix with a Farey interval of $[u/m, (u+m)/(v+n)]$ and multiplication by ES gives one of $[v/n, (u+m)/(v+n)]$: the interval splits at the Farey sum at each bifurcation. Since the matrix one down from the root has a Farey interval of $[0, \frac{1}{2}]$, every matrix below that has its Farey interval within this range, which means that Bézout pair in the second column of each matrix is in fact a winding-number pair.

Examining the tree also justifies our claim that $mv - nu$ is plus or minus one equally often in the tree (save for the first two nodes), since its sign changes each time there is a multiplication by ES rather than S.

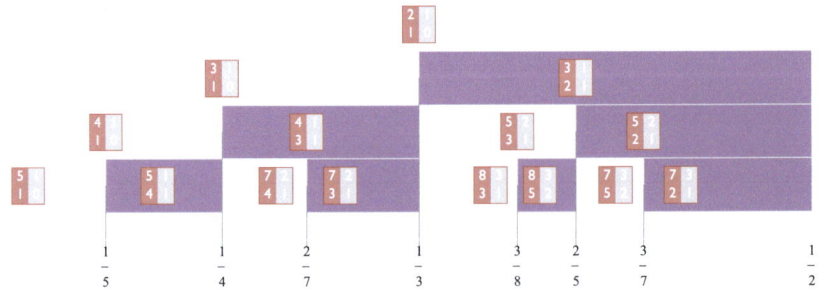

Fig. 3.2 Farey intervals $[u/m, v/n]$ for matrices $\begin{pmatrix} n & m \\ v & u \end{pmatrix}$ coloured blue if the determinant of the matrix is $+1$ and white when it is -1

3.4.1 Fibonacci Pairs

The Fibonacci matrices in Fig. 3.1 are found by consistently choosing the ES branch for each bifurcation from the second node so that, for example

$$\begin{pmatrix} 13 & 5 \\ 8 & 3 \end{pmatrix} = (ES)^4 \cdot E^2 \, S;$$

and recalling that $F_7 = 13$, this generalises easily to

$$\begin{pmatrix} F_{j+1} & F_{j-1} \\ F_j & F_{j-2} \end{pmatrix} = (ES)^{j-4} \cdot E^2 \, S. \tag{3.10}$$

3.5 Prove this.

Together with the previous section this shows that the winding-numbers for a pair of adjacent Fibonacci numbers are the two previous Fibonacci numbers. Moreover we can calculate the associated Farey intervals for use in Chap. 4.

3.6 Show the Farey interval for the Fibonacci pair $(m, n) = (F_j, F_{j+1})$ contains $1/\tau^2 = 1/(1 + \tau)$ but that for $k > 1$ the Farey interval for the generalised Fibonacci pair (F_j^k, F_{j+1}^k) contains $\tau/(1 + k\tau)$.

3.5 Continued Fractions

If we divide each line of the Euclidean algorithm example above by r_i we get

$$\frac{11}{4} - 2 = \frac{3}{4} \tag{3.11}$$

$$\frac{4}{3} - 1 = \frac{1}{3} \tag{3.12}$$

$$\frac{3}{1} - 3 = 0 \tag{3.13}$$

In each case the right hand side fraction is the inverse of the first fraction on the next line and we can solve in reverse order to get

$$\frac{1}{3} = \frac{1}{3} \tag{3.14}$$

$$\frac{3}{4} = \frac{1}{1 + \frac{1}{3}} \tag{3.15}$$

$$\frac{11}{4} = 2 + \frac{1}{1 + \frac{1}{3}} \tag{3.16}$$

So there is a close link between the Euclidean algorithm and the construction of continued fractions. The continued fraction generated by the finite sequence $q_0 \ldots q_N$ is by definition

$$[q_0, \ldots q_N] = q_0 + \cfrac{1}{q_1 + \cfrac{\cdots}{q_{N-1} + \cfrac{1}{q_N}}}. \tag{3.17}$$

When representing a continued fraction less than 1, q_0 will be zero, but otherwise typically the q_i are strictly positive integers. We can usefully relax this rule for the last coefficient:

$$\frac{11}{4} = [2, 1, 3] \tag{3.18}$$
$$= [2, 1 + 1/3]. \tag{3.19}$$

The reason we have used the same label q as in the Euclidean algorithm is that the integer q_is that the Euclidean algorithm generates for the hcf of m and n are exactly the continued fraction coefficients of n/m. Specifically, to compute the continued fraction coefficients $q_0 \ldots q_N$ of a real d we set $\rho_{-1} = d$, $\rho_0 = 1$ and $i = 0$ and then

1. Set an integer $q_i = \lfloor \rho_{i-1}/\rho_i \rfloor$.
2. Set $\rho_{i+1} = \rho_{i-1} - q_i \rho_i$.
3. If $\rho_{i+1} = 0$, set $N = i$ and terminate.
4. Otherwise increment i and repeat.

This is the same as Euclid's algorithm but for the scaled sequence member $\rho_i = r_i/r_{i+1}$. It will terminate if d is rational n/m, because Euclid's algorithm does in that case, but it will not if d is irrational. In either case the intermediate continued fractions generated at each stage represent increasingly good approximations to d.

There is a large literature on continued fractions. Fowler [41] combines the basic results with a relevant historical perspective while Berger [11] adds an informative geometric view.

3.6 Continued Fractions and Möbius Maps

Using the continued-fraction representation, given a set of q_is we can define a function of w as

$$f_{q_0 \ldots q_i}(w) = [q_0 \ldots q_i, w]. \tag{3.20}$$

Taking coefficients from in our running example we have

3.6 Continued Fractions and Möbius Maps

$$f_3 = 3 + \frac{1}{w} = \frac{3w+1}{w+0}$$

$$f_{1,3} = 1 + \frac{1}{3+1/w} = \frac{4w+1}{3w+1} \qquad (3.21)$$

$$f_{2,1,3} = 2 + \frac{1}{1+\frac{1}{3+1/w}} = \frac{11w+3}{4w+1}$$

where we recognise the coefficients of the matrix $M_{2,1,3}$ of Eq. 3.8 arising from the Euclidean algorithm for 11 and 4. Clearly this is not a co-incidence, and by analogy with the previous sections we can see that f can be constructed by repeated function compositions moving us down the tree of Fig. 3.1. Indeed since $f_{q_i} = q_i + 1/w$, the last two can be written as

$$f_{1,3} = f_1(f_3(w)) = f_1 \circ f_3 \qquad (3.22)$$

$$f_{2,1,3} = f_2 \circ f_1 \circ f_3 \qquad (3.23)$$

and in general

$$f_{q_0 q_1 \cdots q_N} = f_{q_0} \circ f_{q_1} \cdots \circ f_{q_N}. \qquad (3.24)$$

Each of the f constructed in this way is a *Möbius map*.

3.6.1 Möbius Maps

Möbius maps have the form[1]

$$f(w) = \frac{aw+b}{cw+d}, \qquad (3.25)$$

and are associated with a coefficient matrix

$$M(f) = \begin{bmatrix} a & b \\ c & d \end{bmatrix}. \qquad (3.26)$$

It can be verified directly that the composition of two Möbius maps is a Möbius map, and that the coefficient matrix of the composition is the conventional matrix product of the coefficient matrices:

---

[1] Poincaré called these Fuchsian functions; they have variously been called automorphic functions, Hilbert modular functions, affine transforms or fractional linear transformations. I follow the authority of Wikipedia, partly because as a native English reader I see an umlaut as conferring scientific respectability.

$$M(f_1 \circ f_2) = M(f_1) \cdot M(f_2). \tag{3.27}$$

One way to see this is to note that

$$f\begin{pmatrix}w_1\\w_2\end{pmatrix} = \frac{w_1}{w_2} \text{ where } \begin{pmatrix}w_1\\w_2\end{pmatrix} = \begin{pmatrix}a & b\\c & d\end{pmatrix}\begin{pmatrix}w_1\\w_2\end{pmatrix}. \tag{3.28}$$

Although we do conventional matrix multiplication to compute the function composition, the functions being composed are not the conventional linear transformations of the real plane $\mathbb{R}^2 \to \mathbb{R}^2$ allowing rotation, scaling and shear. Möbius maps are better thought of as functions on the complex plane $f : \mathbb{C} \to \mathbb{C}$. If w is a complex variable, then because $f(w) = w + b$, $f(w) = aw$, and $f(w) = 1/w$ represent translation, scaling, and inversion in the unit circle, Möbius maps correspond to transformations of the complex plane generated by these geometric tranformations.

In our applications of Möbius maps the coefficient matrix will always have integer entries and moreover have determinant $mv - nu = \pm 1$. Because every such matrix is invertible in integers the maps form a group which I call the modular group.[2] A comprehensive treatment of Möbius transformations is Ford's *Automorphic Functions* [40], but they reappear in many branches of modern mathematics. We don't rely on, but later Chapters will partially rediscover, some well known properties of the modular group. In the language of hyperbolic geometry, when the upper complex half-plane is given the hyperbolic metric, its geodesics are semi-circles (including vertical lines) on the horizontal axis, and the modular group is the symmetry group of these geodesics: f will map axis semi-circles to axis semi-circles [63, 104].

More practically, we can re-cast the computation of 3.21 as a matrix multiplication. Since

$$f_{q_i}(w) = \frac{q_i w + 1}{w + 0} \tag{3.29}$$

the corresponding coefficient matrix is

$$\begin{bmatrix}q_i & 1\\1 & 0\end{bmatrix} = E^{q_i} \cdot S. \tag{3.30}$$

Now the composition

$$f_{q_0 q_1 \cdots q_N} = f_{q_0} \circ f_{q_1} \cdots \circ f_{q_N} \tag{3.31}$$

---

[2] If Wikipedia remains the authority, then the modular group is instead the group of matrices with integer entries and determinant $+1$, and the group with determinants ± 1 is $PSL(2, \mathbb{Z})$; but the same authority also says this is called $SL(2, \mathbb{Z})$ by some authors.

has a coefficient matrix which can be computed as a matrix product

$$M_{q_1q_2\cdots q_N} = (E^{q_1} S) \cdot (E^{q_2} S) \cdots (E^{q_N} S) \begin{pmatrix} 1 & 0 \\ 0 & 1 \end{pmatrix} = \begin{bmatrix} n & v \\ m & u \end{bmatrix}, \quad (3.32)$$

say, as in (3.9). Then

$$f_{q_0\ldots q_N}(w) = \frac{nw + v}{mw + u} \quad (3.33)$$

showing that (3.21) holds true in general.

3.7 Summary

This excursion onto the nursery slopes of hyperbolic geometry and number theory has explored the structure of the Euclidean algorithm as illustrated in Fig. 3.1. In Chap. 4 we will see this structure reappear in the van Iterson diagram for cylindrical lattices.

Part II
Mathematical Theory

Chapter 4
The Geometry of Cylindrical Lattices

Abstract The goal of this chapter is to characterise the shortest vectors in a cylindrical lattice. This will allow us, in the next Chapter, to classify lattices by their shortest vectors in a way that has biological relevance. Cylindrical lattices have a natural representation as 'unrolled' plane lattices in which the cylinder circumference corresponds to a vector of the plane lattice encoding the periodicity; the height of a point in the cylindrical lattice is the component perpendicular to the periodicity vector. Any vector in the lattice has a parastichy number corresponding to the size of this component in units of node height. The principal parastichy vectors are the two shortest vectors in the plane lattice. We show how the the cylindrical spirals formed by these vectors—and the two corresponding parastichy numbers—provide a natural model for biological observations of spiral counts. In order to compute these parastichy numbers in a given lattice we will consider generating pairs of vectors as the building blocks of the lattice, and carry out the main analysis of this chapter: how to deduce from the generating pairs what the lattice parameters are and, especially, what the shortest vectors in the lattice are. The central and suggestive result of this Chapter is Turing's theorem: that the third-shortest vector in a lattice has a parastichy number which is the sum or difference of the two principal parastichy numbers.

4.1 Connections with Botanical Terminology

Figure 4.1 shows a typical, and typically idealised, botanical classification of stem architecture. We will call the number of leaves J at each stem height where they occur the *jugacy*. The first and second patterns are our primary interest in this chapter: they are *monojugate* with a fixed angle between each leaf; in the special case of the distichous ('two-rowed') arrangement the angle is 180°. By contrast the third and fourth patterns in Fig. 4.1 are multijugate. The third, decussate ('crossing') pattern has each succeeding pair at right angles, and when $J > 3$ the arrangement is called a whorled phyllotaxis.

The classification in Fig. 4.1 assumes that there is a clear, and fixed, rise between each occurrence of each whorl but the node numbering assigns only an arbitrary sequence to the nodes at each fixed rise. Yet further patterns are possible: for exam-

30   4 The Geometry of Cylindrical Lattices

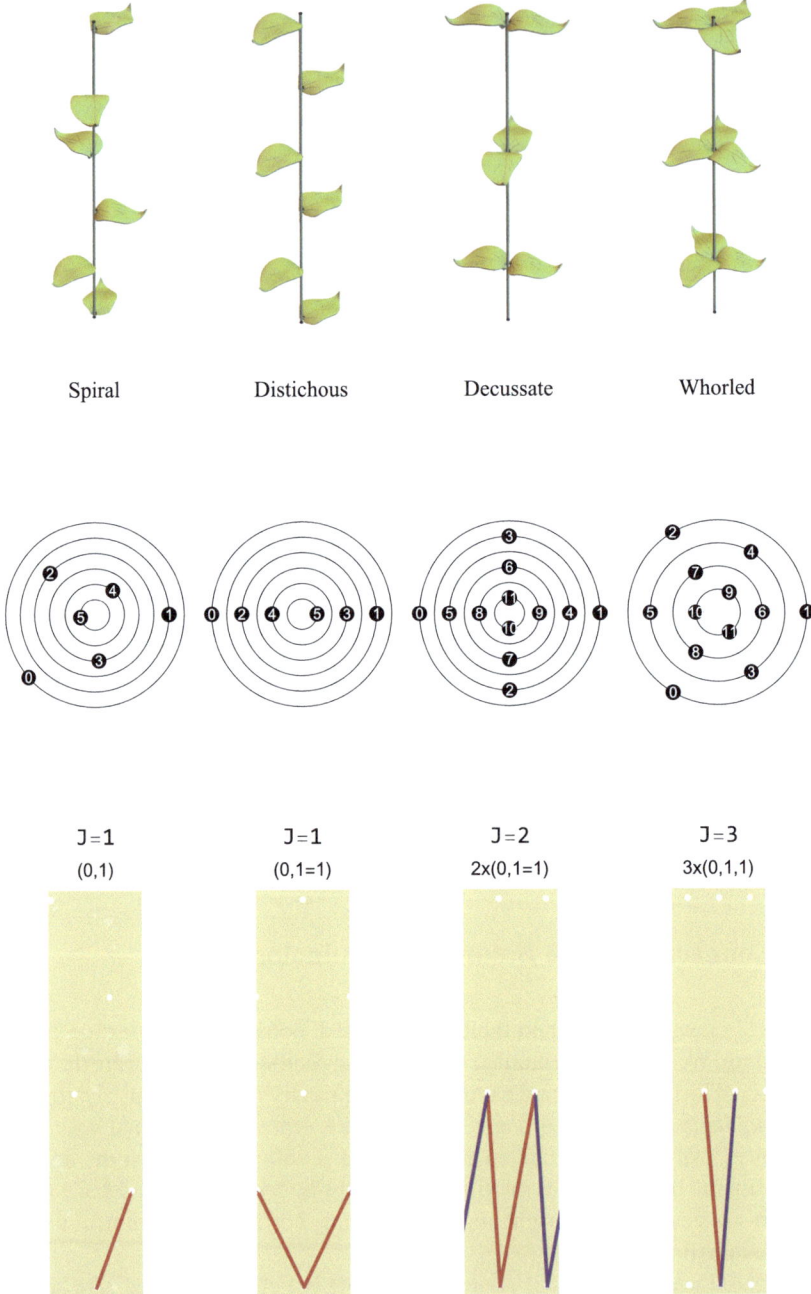

Fig. 4.1 Four different types of leaf arrangements on stems, with jugacy J, a classification by principal parastichy numbers which will be developed in this chapter, and a representation as a lattice on a cylinder, with a point at one height joined to the two nearest points at the next height

ple Yonekura et al. have recognised 'orixate' phyllotaxis in which there is a consistent developmental order within each pair of decussate leaves [134], but we restrict attention to the simpler cases.

These patterns can all be idealised into a family of points on a cylinder, each rotated by a fixed fraction of the circle or *divergence* d from the next and shifted vertically by the *rise*. In the spiral example, the third leaf is roughly vertically above the first one, and the fifth leaf even more so. If there was always a pth leaf directly above the first, after having gone round the stem q times, then all divergences would be rational fractions p/q and this would provide a system of classification of leaf patterns. At its simplest this would come from identifying the *orthostichy leaf* which is the leaf most 'directly' above the first, but this is in many cases impractical or prone to disagreement. An alternative would be to estimate the divergence, which has the attraction that it can be estimated locally from nearby leaf positions and can be averaged over many observations, and then approximating it by a rational fraction. Of course one divergence can be approximated by many different rational fractions so the choice of approximation is another potential source of dispute.

Nevertheless, this classification method is a natural one which works well for small denominators, when the orthostichy leaf is not too far away, and was in successful and widespread throughout the nineteenth century and beyond. But as well as its conceptual shortcomings, it becomes practically inadequate for more complex patterns. Accordingly we will turn to a more modern idea of characterising lattice patterns by their parastichies.

4.2 Cylindrical and Plane Lattices

Cylindrical lattices can be thought of as lattices in the plane with a particular distinguished vector corresponding to the cylindrical periodicity, and in this plane all the normal rules of vector arithmetic hold. Table 4.1 summarises the notation used in this Chapter. We use surface coordinates (x, z) on a cylinder, with a fixed circumference of 1, that extends infinitely in the vertical direction z. At the origin we place a point of the lattice. By rotating an angle $2\pi d$ around the cylinder from the origin and rising by $z = h$ we place a second point. Repetitions of this translation will define the whole lattice, and we will call h the rise and d the divergence.

Given this coordinate system, the kth lattice point, ⓚ, has coordinates $(x, z) = (x_k, kh)$ where $x_k = kd - [kd]$ and $[x]$ is the nearest integer to x, with ties chosen so that $-\frac{1}{2} < x_k \leq \frac{1}{2}$. With this notation the origin is at ⓪. The lattice itself, $\mathcal{L}(d, h)$, is the collection of all such points ⓚ for integer k. See Fig. 4.2. Co-ordinates of lattice points are thus in the cylinder strip $C = (-\frac{1}{2}, \frac{1}{2}] \times \mathbb{R}$, and there is a natural planar lattice which is the unrolled set of points $\mathcal{L} + w(1, 0)$ over all integers w. There is a simple map from the plane back to the cylinder: $(x, z) \to (x - w, z)$ where $w = [x]$ is the *winding number* of the vector (x, z). This maps all the plane lattice points of rise kh onto the unique cylindrical lattice point ⓚ of rise kh.

4 The Geometry of Cylindrical Lattices

Table 4.1 Notation used in this chapter

$[x]$	The nearest integer to x
ⓚ	The kth lattice point, with coordinates $(dk - [dk], kh)$
(x, z)	Real coordinates in a vertical cylinder $[0, 1] \times \mathbb{R}$ or the plane $\mathbb{R}^2$, or more often a vector in $\mathbb{R}^2$ relative to an origin $(0, 0)$. (Warning: at the end of Chap. 5 we use a different $z = d + ih$ with $z \in \mathbb{C}$)
$\mathbf{p}_0$	The vector $(1, 0)$. We call this the *periodicity* vector
$\mathbf{p}_1$	The vector (d, h), usually with $0 < d \le \frac{1}{2}$ and $h > 0$. This is the *generating* vector
$\mathbf{p}_k$	The kth parastichy vector: the vector from ⓪ to ⓚ
$\hat{\mathbf{p}}_k$	The kth complementary vector: the shorter of $\mathbf{p}_k \pm \mathbf{p}_0$
$\mathcal{L}(d, h)$	The lattice on the cylinder generated by the vectors $\mathbf{p}_0$ and $\mathbf{p}_1$: the set $a\mathbf{p}_0 + b\mathbf{p}_1$ for $a, b \in \mathbb{Z}$, after identifying points differing by an integer in x
(m,n) pair	The principal vectors of the lattice: the linearly independent pair $\mathbf{p}_m$ and $\mathbf{p}_n$ (or $\hat{\mathbf{p}}_n$) which are shortest in the lattice). More generally:
(r,s) pair	A generating pair: any pair of vectors that also generate $\mathcal{L}(d, h)$
(m,n) lattice	A lattice in which the principal pair are (m,n)
m, n, u, v	Integers satisfying $mv - nu = \Delta$ and $\Delta = \pm 1$

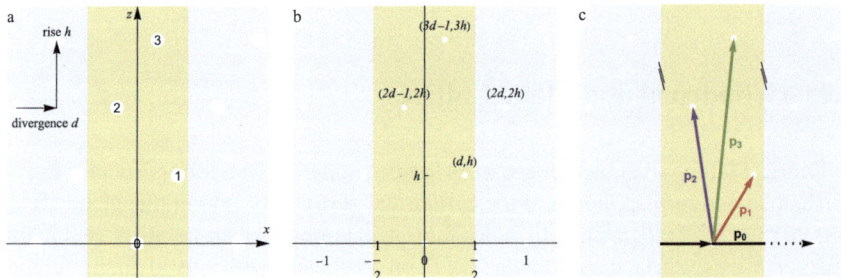

Fig. 4.2 Equivalent parameterisations of the same periodic planar lattice with divergence d and rise h, here $d = 4/10, h = 1/10$. In this case the cylindrical lattice point ③ has co-ordinates $(3d - 1, 3h) = (3d - [3d], 3h)$. In panel (c), some parastichy vectors are shown and $\mathbf{p}_3 = 3\mathbf{p}_1 - \mathbf{p}_0$: the vector $\mathbf{p}_0$ encodes the periodicity of the lattice. In panel (c), the double diagonal lines mark the periodic boundary $x = \pm\frac{1}{2}$ of the cylindrical lattice. Elsewhere in this book the pale green rectangle alone, of width 1, is used as a convention to indicate this periodicity of the cylinder

4.2 Cylindrical and Plane Lattices

More generally, planar lattices are generated from pairs of vectors **a** and **b** as the set of all possible integer sums $u\mathbf{a} + v\mathbf{b}$. It will often happen that two different generating pairs generate the same given lattice. A planar lattice is 1-periodic if it has the vector $(1, 0)$ as a member, and then it has a corresponding cylindrical lattice which is the subset of points with the divergence x in $[\frac{1}{2}, \frac{1}{2}]$.

4.2.1 Labelling Cylindrical Vectors

Each point of the plane lattice corresponds to a vector in $\mathbb{R}^2$ with the normal rules of vector arithmetic. But although the map of planar-lattice points to cylindrical-lattice points is many-to-one, the map of planar vectors to cylindrical vectors is one-to-one, because each pair of points on the cylinder will be joined by a family of vectors, each winding a different number of times around the cylinder, and with co-ordinate components equal to components of the corresponding plane vector. Thus we can do our normal vector arithmetic on the cylinder, although we will be careful about how we label vectors.

First we distinguish one special vector $\mathbf{p}_0 = (1, 0)$ which encodes the periodicity of the lattice. After that, we note that the shortest of all the vectors on the cylinder from ⓪ to Ⓚ is the shortest of the set of vectors from the origin to any of the points in the planar lattice with rise kh; it must be one of the two vectors to the point Ⓚ with winding number zero. We name this shortest of the vectors from the origin to Ⓚ as the *parastichy vector* $\mathbf{p}_k$.

Sometimes adding two parastichy vectors takes the sum outside of the cylinder and then the sum is not a parastichy vector. For the most part we only need to be concerned with parastichy vectors, but for some (mathematically if not biologically) significant special cases, there is a role for the *second* shortest vector to the point Ⓚ. We call this the *complementary vector* $\hat{\mathbf{p}}_k$ (Fig. 4.3). The complementary vector is the shorter of $\mathbf{p}_k \pm \mathbf{p}_0$. There is a special case when $d = \frac{1}{2}$, so that the vectors $(\frac{1}{2}, h)$ and $(-\frac{1}{2}, h)$ are equal-length vectors to the point ①: in this case we label them as $\mathbf{p}_1$ and $\hat{\mathbf{p}}_1$ respectively.

4.1 Show that every parastichy vector has a horizontal component with absolute value less than or equal to $\frac{1}{2}$, and every complementary vector has a horizontal component with absolute value unstrictly between $\frac{1}{2}$ and 1. Show that the sum of two parastichy vectors is either a parastichy vector or a complementary vector.

4.2 Show that it is not true in general that $\mathbf{p}_k = k\mathbf{p}_1$, but that $\mathbf{p}_k = k\mathbf{p}_1 - w_k\mathbf{p}_0$ for some integer winding number w_k. Find a lattice in which there are no integers m, n such that $\mathbf{p}_3 = m\mathbf{p}_1 + n\mathbf{p}_2$.

Integer values of the divergence d will not be very useful, and because $\mathcal{L}(d, h) = \mathcal{L}(1 + d, h)$, the range of d can be taken as $(-\frac{1}{2}, \frac{1}{2})$. The two lattices $\mathcal{L}(d, h)$ and $\mathcal{L}(-d, -h)$ are the same and we will assume that $h > 0$. By contrast, the lattices $\mathcal{L}(d, h)$ and $\mathcal{L}(-d, h)$ are mirror-images of each other.

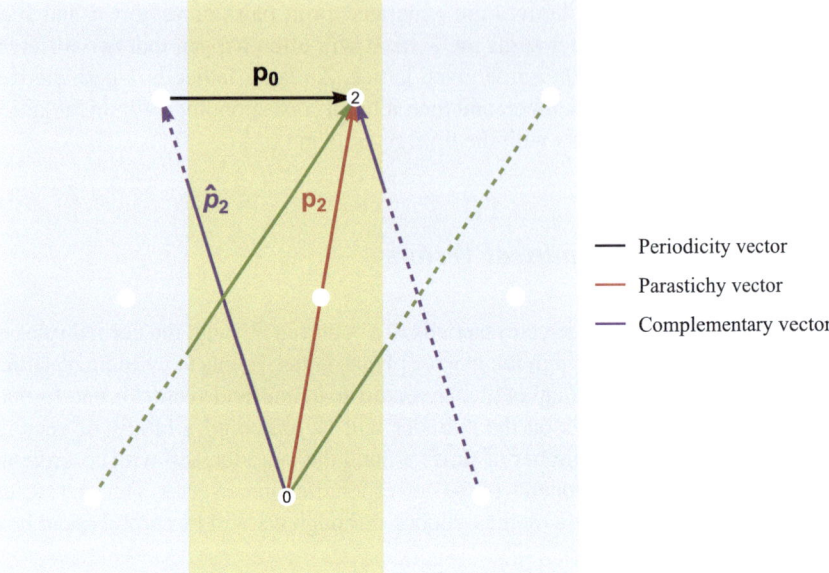

Fig. 4.3 The parastichy vector $\mathbf{p}_2$, in red, is the shortest vector on the cylinder to the point ② and always has $|x_2| \leq \frac{1}{2}$. The second-shortest vector to the point will wind in the opposite direction and is called the complementary vector $\hat{\mathbf{p}}_2$, in blue. Here $d = 17/72$ so $x_2 = 2d$ and $\hat{\mathbf{p}}_2 = \mathbf{p}_2 - \mathbf{p}_0$ where $\mathbf{p}_0 = (1, 0)$ is the horizontal vector around the circumference of the cylinder. A further, un-named vector is shown in green that winds more than a full turn about the cylinder: there are infinitely many other vectors from periodic images of the point ⓪ to the point ②, each with a non-zero winding numbers

4.3 Parastichies

The idea behind parastichies is to characterise a lattice not by its divergence, nor by identifying one point nearly above the origin, but by the most visible structures in the lattice, the families of straight lines that the eye naturally constructs through points in the lattice (Fig. 4.4).

We model this psychological choice through the parastichy vector. For an integer m, the origin-parastichy of order m is the infinite line on the cylinder winding through ⓪ and Ⓜ and with slope mh/x_m (or zero in the case $m = 0$). This choice of slope is equivalent to choosing the line that traverses the smallest x distance between 0 and Ⓜ, and the portion of this line between 0 and Ⓜ coincides with the vector $\mathbf{p}_m = (x_m, mh)$ defined above as the parastichy vector.

For integer m, we define the m-foliation as the family of m lines on the cylinder containing the origin-parastichy of order m and the parallel lines to it through the points ①, ..., (m-1). (These m lines might not be distinct: there are only 5 distinct members of the 10-foliation in Fig. 4.4.) An m-parastichy is any one member of the

4.3 Parastichies

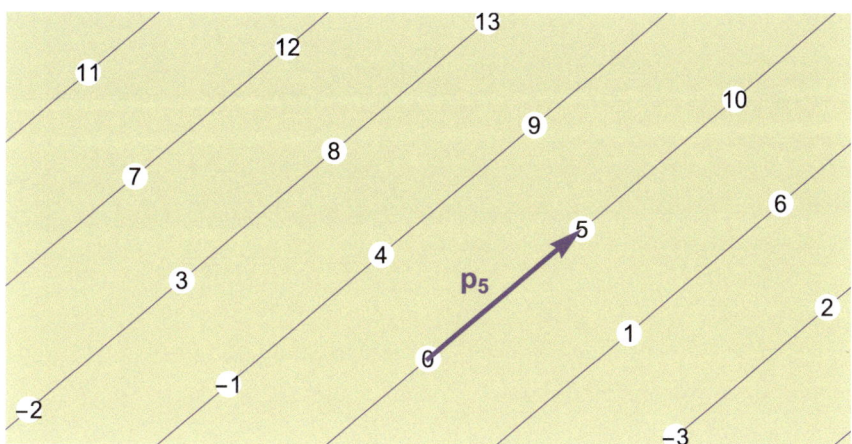

Fig. 4.4 The family of 5-parastichies is a set of five lines on the cylinder, each member passing through one of the points ⓪ to ④. The line that passes through the origin ⓪, by definition the origin-parastichy of order 5, also passes through the points ⑤ and all points ⑤ₖ. The parastichy vector $\mathbf{p}_5$ (arrowed) is the shortest vector from ⓪ to ⑤. This lattice has divergence $d = 17/72$ and rise $h = 0.03$

m-foliation, though usually we are thinking of the origin-parastichy of order m. The 1-parastichy, which passes through every point of the lattice, is known as the *genetic spiral*. and each horizontal line through a point of the lattice is a 0-parastichy.

Most of the time the (e.g.) 5-parastichy, the point ⑤, and the vector $\mathbf{p}_5$, can be thought of interchangeably as standing for each other; but $\hat{\mathbf{p}}_5$ is a vector winding the other way around the cylinder to $\mathbf{p}_5$ although it also arrives at ⑤.

4.3.1 Visible Points and Parastichies

In Fig. 4.4 the point ⑩ is not 'visible' from the origin because the point ⑤ can be thought of as obscuring it; the origin-parastichies of order 5 and 10 are the same lines. We say that ⓜ is a *visible point*, or the m-parastichy is a *visible parastichy*, if the latter is linearly independent of all the n parastichies for all $|n| < |m|$.

4.3 Check visually that in the lattice of Fig. 4.4 the 7- and 9-parastichies are visible but the 2 parastichy is not. If a point ⓝ is on an m-parastichy then show that ⓜ₊ₙ is on the same one.

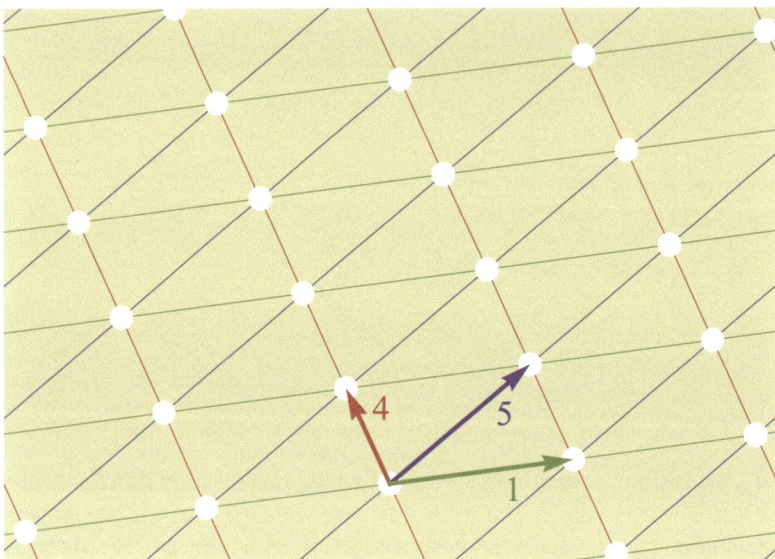

Fig. 4.5 Some of the more obvious parastichies and their parastichy vectors in a lattice with $d = 17/72$ and $h = 0.03$. The blue lines highlight the family of 4-parastichies, the red lines the family of 5-parastichies, and the green ones the family of 1-parastichies. There are four members of the 4-parastichies, and the member that passes through the point 0 also passes through the point 4. For this lattice, (1,4) and (4,5) are each examples of a generating and opposed pair: in each case the two parastichies wind in opposite directions, the parallelogram defined by the pair tiles the lattice and every lattice point is at a vertex of one of the parallelograms; the edges of the parallelograms form the parastichy lines

4.4 Parastichy Pairs

So far we have formalised what it means to be able to 'see' a straight line in the lattice as a visible parastichy. We now concentrate on pairs of parastichies, as it is these pairs we will use to classify our lattices. Figure 4.5 shows two examples of parastichy pairs which might naturally be identified in a lattice.

Definition 4.1 The (m,n)-parastichy pair is the pair of parastichy vectors $\mathbf{p}_m$ and $\mathbf{p}_n$. The (m,n̂)-complementary pair is the pair of the parastichy vector $\mathbf{p}_m$ and the complementary vector $\hat{\mathbf{p}}_n$.

Complementary pairs are only really needed for thinking about the special cases (1,k̂) and do not play a role in large Fibonacci lattices. The most important of these special cases, though is frequently seen. Plants may show *distichous* phyllotaxis, with successive organs alternating in front and behind with a divergence of 180°, correspond to a lattice with a (1,1̂) parastichy pair as in Fig. 4.1. The hat can and will be dropped from the second **1** when it is clear which of the vectors is meant.

4.4.1 Opposed Parastichy Pairs

A parastichy pair (m,n) is *opposed* if m/x_m and n/x_n have opposite sign, where x_m and x_n are the x-components of $\mathbf{p}_m$ and $\mathbf{p}_n$. A complementary pair (m,m̂) is always opposed. Historically, the idea of opposed pairs, with parastichies that wind in opposite directions, was an important organising idea for lattice classification, and they do play a role, but here we put more emphasis on the idea of a generating pair, and, later, even more on the principal pair.

4.5 Generating Pairs

There are many pairs of visible parastichies in a lattice, but Fig. 4.6 shows that finding one is not enough to characterise our lattices. Looking at the lower left diagram of the Figure, both $\mathbf{p}_5$ and $\mathbf{p}_7$ are visible but knowing this pair exists doesn't tell us about $\mathbf{p}_4$: even though the pair are linearly independent, we can't always express other parastichy vectors as their integer sums. To make that precise we use the idea of a generating pair.

Definition 4.2 A pair of vectors **a** and **b** is *generating* in a lattice $\mathcal{L}$ if every vector of $\mathcal{L}$ can be expressed as a vector sum $q\mathbf{a} + r\mathbf{b}$ for integer q, r.

Most of the time the pair of vectors we see making up a generating pair will themselves each be parastichy vectors, but that is not required by the definition. If a pair is generating in the lattice $\mathcal{L}(d, h)$ it is also generating in any other lattice with the same divergence, independently of h. In general one lattice has infinitely many generating pairs.

4.4 Show this.

Since every point in our lattice has a rise which is an integer multiple of h, a necessary condition for a pair to be generating is that it can generate all these h multiples in the vertical component of its sums:

Theorem 4.1 *If a parastichy pair (m,n) is generating, then m and n are co-prime.*

Proof Write $\mathbf{p}_1 = v\mathbf{p}_m - u\mathbf{p}_n$; then the rise of $\mathbf{p}_1$ shows $mv - nu = 1$ holds and Sect. 2.2 shows that $|m|, |n|$, and hence m, n, are co-prime. The converse is not true: see Fig. 4.6. In particular though 1 and 1 are co-prime the parastichy pair (1,1) is not generating. □

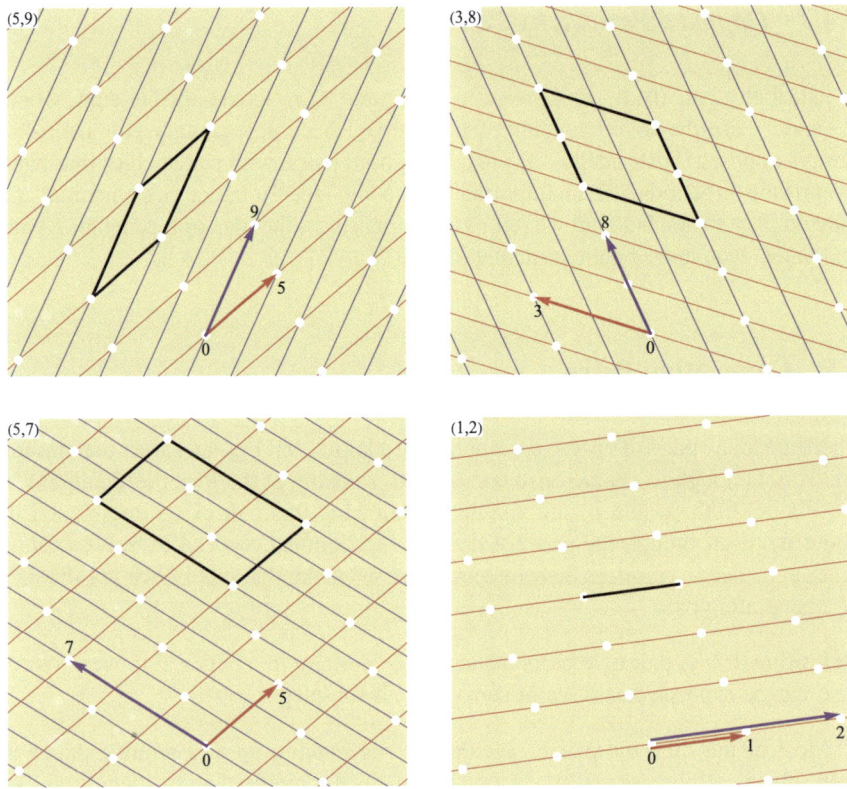

Fig. 4.6 Further parastichy pairs in the same lattice as Fig. 4.5. **a** (5,9) is a generating but not opposed pair and the corresponding parallelogram (in black) has no lattice points except at its corners; **b** (3,8), though a linearly independent pair, is nonopposed and nongenerating pair; **c** although $\mathbf{p}_5$ and $\mathbf{p}_7$ are both visible parastichies, and are linearly independent, (5,7) is not a generating pair because there are lattice points internal to its parallelogram; **d** (1,2) is not a linearly independent pair and cannot be generating, although for this spiral lattice, (1,$\hat{2}$) is a generating pair

This is not a sufficient condition. For example:

Theorem 4.2 *If $\mathbf{p}_m$ and $\mathbf{p}_n$ are co-linear then they are not generating.*

Proof Co-linearity means that any vector sum of $\mathbf{p}_m$ and $\mathbf{p}_n$ lies on the line in the plane through ⓜ and ⓝ. For some large enough k this will contain a point of the plane lattice of rise kh with co-ordinates outside of the cylinder. The vector to this point is therefore not a parastichy vector but is the only candidate for a vector sum with rise kh, and so the parastichy vector $\mathbf{p}_k$ is not an integer sum of $\mathbf{p}_m$ and $\mathbf{p}_n$. □

Excluding co-linearity is still not enough, as part (b) of Fig. 4.6 suggests. The next section finds an algebraic condition for a pair to be generating.

4.5 Show that (0,1) is always generating, and that (0,n) never is for $n > 1$.

4.5 Generating Pairs

Fig. 4.7 The pair of red and blue vectors $\mathbf{p}_3$ and $\mathbf{p}_2$, and the pair of white and green vectors $\mathbf{p}_0$ and $\mathbf{p}_1$ each generate the same lattice

4.6 Show that the complementary pair $(\mathsf{m},\hat{\mathsf{m}})$ is generating only when $|m| = 1$.

4.7 Show that in the lattice $\mathcal{L}(d = 17/72)$, $\mathbf{p}_4$ cannot be expressed as an integer sum of $\mathbf{p}_5$ and $\mathbf{p}_7$ and thus that $(5,7)$ is not generating in this lattice.

4.5.1 Generating Pairs as Basis Vectors

A helpful way to think of the generating pair (m,n) is that $\mathbf{p}_m$ and $\mathbf{p}_n$ provide a coordinate basis for the plane, and the lattice points are the ones with integer coordinates in that basis. From this perspective, a cylindrical lattice is not generated merely by the first parastichy vector $\mathbf{p}_1$, but instead by the pair comprising $\mathbf{p}_1$ together with the periodicity vector $\mathbf{p}_0$, which is used as necessary to pull multiples of $\mathbf{p}_1$ back into the cylindrical strip (Fig. 4.7). Indeed we saw in Exercise 4.5 that $\mathbf{p}_0$ and $\mathbf{p}_1$ are always generating, or in this sense always form a basis pair for the cylindrical lattice. But there are many possible generating pairs. One of the ways that this is viewpoint is useful that it gives us a way to characterise generating pairs by thinking of them as corresponding to a different choice of basis for the space of lattice vectors in the plane and produce the central algebraic result of this chapter:

Theorem 4.3 *In any lattice $\mathcal{L}(d, h)$, the following are equivalent for coprime integers m, n:*

1. *The pair of parastichy vectors $\mathbf{p}_m$ and $\mathbf{p}_n$ is a generating pair.*
2. *The area of the parallelogram formed by $\mathbf{p}_m$ and $\mathbf{p}_n$, namely the vector cross product $h\Delta_{mn} = \mathbf{p}_m \times \mathbf{p}_n$, satisfies $|\Delta_{mn}| = 1$.*
3. *The integers $u = [md], v = [nd]$ satisfy $|mv - un| = 1$ and the parastichy vectors satisfy*

$$\begin{pmatrix} \mathbf{p}_n \\ \mathbf{p}_m \end{pmatrix} = \begin{pmatrix} n & -v \\ m & -u \end{pmatrix} (\mathbf{p}_1 0). \tag{4.1}$$

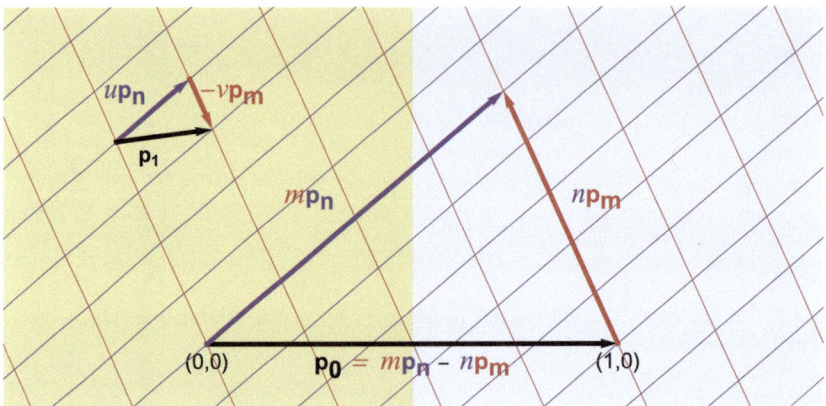

Fig. 4.8 A generating pair must by definition be able to express other lattice vectors, and in particular $\mathbf{p}_0$ and $\mathbf{p}_1$. Theorem 4.3 finds the necessary coefficients. Here $d = 17/72$, and the $(m,n) = (5,4)$ pair is generating, with $x_m = 13/72$, $x_n = -4/72$, $u = 1$, $v = 1$, $\Delta = mv - nu = 1$ The larger triangle is formed by translating $\mathbf{p}_m$ by $\mathbf{p}_0$, and then scaling it by n until it meets $m\mathbf{p}_n$; that must be a lattice point if the pair is generating. This illustrates $\mathbf{p}_0 = m\mathbf{p}_n - n\mathbf{p}_m$. In the smaller triangle, out of every u and v satisfying $mv - nu = 1$, and thus giving $v\mathbf{p}_m - u\mathbf{p}_n$ rise 1, the same as $\mathbf{p}_1$, there is exactly one (u, v) pair giving a parastichy vector which is therefore $\mathbf{p}_1$

Proof Suppose (1) is true. Since the pair $\mathbf{p}_m$ and $\mathbf{p}_n$, is generating the lattice is tiled by the $(\mathbf{p}_m, \mathbf{p}_n)$ parallelogram with an area per point of $h|\Delta_{mn}|$. But the $(\mathbf{p}_0, \mathbf{p}_1)$ parallelogram also tiles the same lattice with an area per point of $h = (1, 0) \times (d, h) = h\Delta_{01}$. These areas must be the same, so that $|\Delta_{mn}| = |\Delta_{01}| = 1$, and (1) $\implies$ (2).

By definition $\mathbf{p}_m = m\mathbf{p}_1 - [md]\mathbf{p}_0$, $\mathbf{p}_n = n\mathbf{p}_1 - [nd]\mathbf{p}_0$. Assume (2) is true and set $u = [md]$ and $v = [nd]$, so (4.1) holds and the cross product $h\Delta_{mn} = \mathbf{p}_m \times \mathbf{p}_n = (mv - un)(\mathbf{p}_1 \times \mathbf{p}_0)$ gives $|mv - un| = 1$ and so (2) $\implies$ (3).

Suppose (3) holds, and invert equation (4.1) in integers to see

$$\begin{pmatrix}\mathbf{p}_1\\\mathbf{p}_0\end{pmatrix} = \pm 1 \cdot \begin{pmatrix}-u & v\\-m & n\end{pmatrix}\begin{pmatrix}\mathbf{p}_n\\\mathbf{p}_m\end{pmatrix}. \quad (4.2)$$

This shows $\mathbf{p}_0$ and $\mathbf{p}_1$ are expressible as integer sums of $\mathbf{p}_m$ and $\mathbf{p}_n$ and so the pair $\mathbf{p}_m$ and $\mathbf{p}_n$ is generating and (3) $\implies$ (1). □

Figure 4.8 illustrates the properties $m\mathbf{p}_n - n\mathbf{p}_m = \pm\mathbf{p}_0$ and $v\mathbf{p}_m - u\mathbf{p}_n = \pm\mathbf{p}_1$ shown by (4.2).

The parallelogram tiling of Theorem 4.3 also shows the following, since the m and n parastichies are formed by the boundary of the $\mathbf{p}_m, \mathbf{p}_n$ parallelogram.

Theorem 4.4 *If the parastichy pair (m,n) is generating then the plane parastichies of order m and n intersect only at points of the lattice in the plane.*

The converse of the Theorem is not true; the intersections of the 3- and 8-parastichies in Fig. 4.6 are exactly the points of $\mathcal{L}$ but (3,8) is not generating for that lattice. This Theorem allows a more algebraic demonstration of the requirement that $|\Delta| = 1$ for a generating pair, by calculating the coordinates of the apex point in Fig. 4.8: this has rise mnh/Δ_{mn} which must be an integer multiple of mn since it is a lattice point and so $|\Delta_{mn}| = 1$ since m and n are co-prime.

4.8 If (m,n) is generating, then $\mathbf{p}_m$ lies on the adjacent n-parastichy to $\mathbf{p}_n$.

Theorem 4.3 is not as general as it could be because not all generating pairs are pairs of parastichy vectors. We will also encounter the case of the pair of vectors $\mathbf{p}_1$ and $\hat{\mathbf{p}}_n$ which are a parastichy vector and a complementary vector. But if need be, each part of Theorem 4.3 can be translated as a statement about $\hat{\mathbf{p}}_n = \mathbf{p}_n \pm \mathbf{p}_0$ instead of $\mathbf{p}_n$.

4.9 Extend the definition of Δ to complementary vectors. Find Δ_{1n} and $\Delta_{1\hat{n}}$. Find d-intervals on which $\Delta_{12} = 1$ or $\Delta_{1\hat{2}} = 1$. Rewrite (4.1) for the case $\hat{\mathbf{p}}_n = \mathbf{p}_n + \sigma \mathbf{p}_1$.

Theorem 4.3 gives us a criterion, $|\Delta_{mn}| = 1$, for a parastichy pair to be generating which is independent of h: whether a parastichy pair is generating or not is independent of the rise, as can be seen by scaling the diagrams of Fig. 4.6 vertically. Note also that (m,n) are not claimed to be unique: for fixed d and thus $\mathbf{p}_0$ and $\mathbf{p}_1$, every integer matrix with determinant of modulus 1 in (4.1) yields a different generating pair.

4.6 Estimating the Divergence for a Generating Parastichy Pair

Suppose that we observe a lattice with a generating parastichy pair of (55,89). Although this is compatible with a range of divergences d, they turn out all to be in a small interval around the Fibonacci angle. To make this precise, we will compute *generating intervals*. Since, as we have just seen, whether or not a pair are generating in a lattice $\mathcal{L}(d, h)$ is independent of h, these intervals only depend on d:

Definition 4.3 The generating interval for the integers m and n is the subset of the d-interval [0, 1] for which (m,n) is generating in $\mathcal{L}(d, \cdot)$.

It's apparent from Theorem 4.3 that the generating interval is exactly when $|\Delta = mv - nu| = 1$. This integer valued function is dependent on the real variable d, so it must be a step-function of d. A few special cases, like $\Delta_{01} = 1$, and $\Delta_{1\hat{1}} = 1$ are constant on the whole d-interval, but in general the work in calculating Δ consists in finding where these step changes are.

For simple parastichy numbers we can calculate generating intervals directly, along with the complementary-generating interval for the integers m and n, which is the subset of [0, 1] for which (m,n̂) is generating in $\mathcal{L}(d, \cdot)$.

4.10 Find the generating and complementary-generating intervals for $m = 0, n = 1$ and for $m = 1, n$.

These direct calculations become more irksome for larger numbers, but we can also calculate generating intervals using winding-number pairs. The strategy to find the generating interval for m, n is first to find u, v satisfying equation $mv - nu = \pm 1$, so we can satisfy Theorem 4.3 and then to solve the piecewise linear equations $[md] = u$ and $[nd] = v$ for d. For a general one of these Bézout pairs, the resulting interval won't be in $[0, \frac{1}{2}]$, although there will be one related by an integer difference or the mirror symmetry $d \to 1 - d$. In fact, the selection of the winding-number pair from the possible Bézout pairs in Sect. 3.2 was so as to ensure we get a generating interval in $[0, \frac{1}{2}]$, as we will see in the proof of Theorem 4.5.

4.11 Show that the d-intervals on which $\Delta_{mn} = \pm 1$ are related by the mirror symmetry.

Example 4.12 shows the relationship between Fibonacci generating pairs and the golden ratio.

4.12 Show that pairs of large enough adjacent Fibonacci numbers are parastichy numbers which are generating in a lattice only if the divergence in $[0, \frac{1}{2}]$ is near to $1/\tau^2$.

4.7 Opposed Intervals

Similarly to calculating generating intervals, we can calculate opposed intervals: those on which a given parastichy pair is opposed. There are no more than two disjoint generating intervals for a given pair, related by the mirror symmetry, but there can be many opposed intervals. We recall $x_m = md - [md]$, $x_n = nd - [nd]$. For m, n positive integers not both equal to one, the pair (m,n) is opposed if $x_m x_n < 0$. We define (0,1) and (m,m̂) to be opposed.

4.8 The Fundamental Theorem of Phyllotaxis

Finally, we can take the intersection of the generating interval and the opposed interval to find the generating and opposed interval. This is the d–interval on which the pair of vectors (m,n) is both generating and opposed in the lattice $\mathcal{L}(d, h)$. When we calculate this we get what Jean called the 'Fundamental Theorem' of Phyllotaxis [60]:

Theorem 4.5 *Suppose that a lattice $\mathcal{L}(d, h)$ is generated by d in $[0, \frac{1}{2}]$ and that $1 < m < n$. If the parastichy vectors $\mathbf{p}_n$ and $\mathbf{p}_m$ are generating and opposed, and u and v are the winding-number pair for m and n defined in Sect. 3.2, then d is in the Farey interval $G = [u/m, v/n]$.*

4.8 The Fundamental Theorem of Phyllotaxis

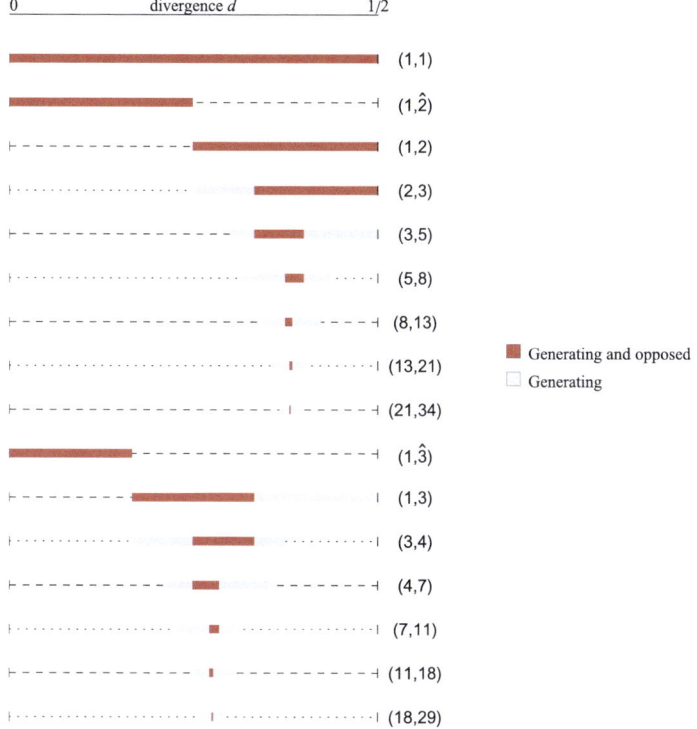

Fig. 4.9 Generating intervals for a range of (m,n) parastichy pairs. The horizontal line represents the d interval $[0, \frac{1}{2}]$. It is drawn dashed if it is the $\Delta_{mn} = +1$ generating interval which falls into $[0, \frac{1}{2}]$ and dotted if it is the $\Delta_{mn} = -1$ interval. Red bars show the subinterval on which the (m,n) pair is generating and opposed; grey bars show the subinterval on which it is generating and unopposed

This says that if we see m spirals one way and n the other, the divergence of the lattice will be in the range $[u/m, v/n]$. The winding-number pair, which already played a discreet role in Theorem 4.3, appears again in the Fundamental Theorem, without a very satisfying reason. The proof below doesn't add much explanation, but one will appear in the next Chapter.

Proof Temporarily allow u and v to be any Bézout pair for m and n, and let M be the d-interval on which $[md] = u$ and N that where $[nd] = v$. The third part of Theorem 4.3 says that the lattice $\mathcal{L}(d, h)$ has $\mathbf{p}_m$ and $\mathbf{p}_n$ as a generating pair on the interval $M \cap N$.

From $[md] = u$ etc. we can calculate the endpoints of $M = [L_m, R_m]$ and $N = [L_n, R_n]$ as $mL_m = u - \frac{1}{2}, mR_m = u + \frac{1}{2}, nL_n = v - \frac{1}{2}, nR_n = v + \frac{1}{2}$. Setting $\Delta = nu - mv$, noting that $\Delta = -1$ if $u/m < v/n$ and $\Delta = +1$ if $u/m > v/n$ we find

$$2mn(u/m - L_m) = n$$
$$2mn(v/n - L_m) = n - 2\Delta$$
$$2mn(u/m - L_n) = m + 2\Delta$$
$$2mn(v/n - L_n) = m$$

so that with either choice of $\Delta = \pm 1$, both L_m and L_n are less than or equal to the smaller of u/m and v/n. A similar argument shows that both R_m and R_n are larger than the greater of u/m and v/n. so the Farey interval $G = [u/m, v/n]$ is contained within the interval $M \cap N$.

On the interval $M \cap N$, the horizontal components of the parastichy vectors are $x_m(d) = md - u$ and $x_n(d) = nd - v$. These are increasing functions with unique zeros at u/m and v/n respectively; both horizontal components are negative at the beginning of $M \cap N$ and both are positive at the end and change sign exactly one, at u/m and v/n respectively, so that the Farey interval G is the only subinterval of $M \cap N$ on which the two parastichy vectors are opposed, and so is a generating and opposed interval.

Recall from Sect. 3.2 that the winding-number pair u, v for m, n has the property that their Farey interval $G = [u/m, v/n] \subseteq [0, \frac{1}{2}]$. If instead u, v were a different Bézout pair for m, n, then they are either of the form $u' = m - u$, $v' = n - v$, which would lead to a Farey interval of $G' = 1 - G$ and so outside $[0, \frac{1}{2}]$, or $u' = u + km$, $v' = v + kn$ which would lead to a Farey interval $G' = [u/m + k, v/n + k]$ outside $[0, 1]$.

So the generating and opposed interval in $[0, \frac{1}{2}]$ is exactly the Farey interval for the winding-number pair. □

Jean's statement of this theorem [60] is incorrect in the case $m = 1$. This is because of the complication which arises in the following exercise, but although of historical significance[1] there is little deeper importance to this edge case and it is excluded in my version of the Theorem above.

4.13 Calculate generating and opposed intervals for the integer pairs $m = 1, n = 1$ and $m = 1, n > 1$.

Like Theorem 4.3, this Theorem 4.5 is independent of the rise, and does not claim that the generating and opposed pair is unique for d. Figure 4.9 gives examples of generating opposed intervals.

As Jean's name for it suggests, this Theorem has historically been given some prominence. One reason is the following special case when the parastichy numbers are adjacent Fibonacci numbers:

---

[1] Trying to understand these complications is what led me to write this half of this book, and in particular to introduce the complementary vector.

Theorem 4.6 *Lattice divergences which lead to Fibonacci structure are very close to the golden angle:*

1. *If F_j and F_{j+1} are successive Fibonacci numbers larger than 1 then the interval on which (F_j, F_{j+1}) is both generating and opposed in $[0, \frac{1}{2}]$ contains the point $1/\tau^2$ and has a width which shrinks as fast as τ^{-2j}.*
2. *If F_m^k and F_{m+1}^k are successive members of the sequence $F_0^k = 1$, $F_1^k = k$, $F_{i+1}^k = F_i^k + F_{i-1}^k$ for integer $k > 2$, then the interval on which (F_m^k, F_{m+1}^k) is generating and opposed in $[0, \frac{1}{2}]$ contains the point $\tau/(1+k\tau)$ and shrinks as fast as τ^{-2j}.*

Proof We have from Sect. 3.4.1 that $F_j^k F_j - F_{j+1}^k F_{j-1} = (-1)^{j-1}$, so that we can set $m = F_j^k$, $n = F_{j+1}^k$, $u = F_{j-1}$, $v = F_j$ and have $mv - un = (-1)^{j-1}$ and $0 < v < n$, and $0 < u < m$. So the interval $G = [u/m, v/n]$ of the theorem is $[F_{j-1}/F_j^k, F_j/F_{j+1}^k]$. Moreover it has width $1/F_j^k F_{j+1}^k$ which is of order τ^{-2j}. For $k > 2$ the interval is already in $[0, \frac{1}{2}]$; for the Fibonacci numbers, with $k = 1$, u/m and v/n are both larger than a half and the interval G is close to $1/(1+\tau)$ so its mirror interval in $[0, \frac{1}{2}]$ is $1 - G$ containing $1/\tau^2$. □

4.14 Compute the divergence interval in $[0, \frac{1}{2}]$ on which (55,89) is a generating opposed pair.

So we can apparently deduce from the fact of 55 and 89 parastichies in Fig. 1.2 that the arrangement of successive florets in the sunflower head is at an angle very close to the golden angle Φ. However recall from Exercise 4.12 that this also follows from the fact that the Fibonacci parastichy pair is generating: observing that it is opposed merely tightens the interval around Φ. But more fundamentally we also need to assume, incorrectly, that the seeds are indeed arranged in a lattice in the first place: a more accurate description of the Theorem is that it says what the divergence must be *if* the arrangement was a lattice.

4.9 Principal Parastichies

One lattice can have many generating, or generating and opposed, parastichy pairs. We want to pick a principal pair that preserves the idea of being the 'most obvious' to the eye. One strong idea of obviousness comes from returning to the motivating example of the sunflower head. In practice each seed is typically in direct contact with four others and shaped into a corresponding diamond. It is these seeds which are directly adjacent which define, to the eye, the spirals. In the language of our mathematical lattice it is the parastichies through the points nearest to the origin.

Definition 4.4 The principal vector (or first principal vector, or principal parastichy) of the lattice is the parastichy vector of non-negative rise which is shortest. Ties will be common, and if we must we choose the vector with the largest x component. If the first principal vector is $\mathbf{p}_m$ then the first parastichy number is m. The second principal vector is the shortest lattice vector which is not co-linear with the principal vector. The nth principal vector is the nth shortest lattice vector not co-linear with the first to $(n-1)$th principal-vectors.

The first principal vector has to be a parastichy vector, and the second principal vector will normally be a parastichy vector or sometimes a complementary vector. The principal parastichy pair of vectors, together with the angle between them, is enough to completely characterise any monojugate lattice, and we will use the corresponding parastichy numbers as a label for the lattice:

Definition 4.5 An (m, n) lattice is a lattice $\mathcal{L}(h, d)$ in which the first and second principal vectors have parastichy numbers m and n respectively. [2]

Similarly a (m, n, t) lattice is a lattice in which the first second and third parastichy numbers are m, n and t. If the first and second principal vectors are equal in length we alternatively describe the lattice as an (m = n,t) lattice. Principal pairs are always generating pairs and a proof of this will become apparent after we have seen how to construct principal pairs.

Whether a parastichy vector is a principal one is an h-dependent property, unlike the properties of being opposed or generating which are h-independent. At any fixed d, for h large enough, the principal vector is always $\mathbf{p}_0$; if $d = 1/\tau^2$, for example, we will see how the principal parastichy number increases through the Fibonacci numbers as h decreases. We will see below that lattices with ties in length between principal parastichy vectors, though non-generic, are particularly important in organising the classification, and that hexagonal lattices where the first three parastichy vectors are equal in length play a crucial role. Figure 4.10 shows principal parastichy pairs for the special case of $d = 0$ and distichous patterns $d = \frac{1}{2}$. In the Figure, we have stretched the ((m, n)) notation a little further to label the lattices. Now, for example, the (1=$\hat{1}$,2) lattice is one in which the parastichy vectors $\mathbf{p}_1$ and the complementary vector $\hat{\mathbf{p}}_1$ tie as the principal parastichy vector and $\mathbf{p}_2$ is the third-shortest parastichy vector.

4.10 Opposed and Non-opposed Lattices

An opposed (or non-opposed) lattice is one in which the principal parastichy pair is an opposed (or non-opposed) pair. Non-opposed lattices are rare but mathematically possible and crucial in understanding the bifurcation diagrams of the next Chapter: we will construct an example of a non-opposed lattice in Fig. 5.3.

---

[2] We are using the same (m, n) typeface in Definitions 4.5 and 4.1: the lattice is being labelled by the label of its principal parastichy pair. Later we will use the same typeface for the empirical pairs of parastichy counts which this models.

4.11 Spiral Lattices

Fig. 4.10 Lattices, and principal parastichies, in the special cases $d = 0$ and distichous patterns $d = \frac{1}{2}$, and varying h. A circle equal in diameter to the shortest principal parastichy vector is drawn around each point. Red lines correspond to the principal parastichy, and blue lines to the parastichy which has the next strictly shorter parastichy vector

4.11 Spiral Lattices

Before we go on to the main result we can look at a special case which is algebraically simple, although not applicable to large Fibonacci number phyllotaxis. This is the *spiral lattice* which by definition is a lattice $\mathcal{L}(d < \frac{1}{4}, h)$ as in the examples in Fig. 12.1. These are lattices in which the genetic spiral or 1-parastichy visually dominates, and in which the idea of a principal pair is not the most observationally important. Nevertheless spiral lattices are the first initial cases that will begin our

Fibonacci framework, and calculating principal pairs directly for such lattices can help explain the marginal structure of the diagrams in the following Chapter.

4.15 Fix $d < \frac{1}{4}$ and set $k = \lfloor 1/2d \rfloor - 1$. Show that as h decreases, the principal parastichy pair of the lattice $\mathcal{L}(d, h)$ begins as (0,1) and then passes through (1,0), (1,$\hat{1}$), (1,$\hat{2}$),...., (1,$\hat{k}$), until reaching (1,k+1) and that both principal vectors are parastichy vectors for all h values smaller than this.

4.12 Turing-Euclid Reduction of Generating Pairs

The examples of the previous section explicitly computed the principal vector pairs for the special case of spiral lattices, and now we turn to calculating them in general. The principal vectors can often be found by eye—indeed that is the point of them—and certainly always by an exhaustive search of what is only a finite number of possibilities, since they must be at least as short as $\mathbf{p}_0$ and $\mathbf{p}_1$, which bounds the possible rise. But they can also be found by a reduction process from any vectors which generate the lattice. One reason to do this is in the proof of Turing's Theorem below. Moreover this reduction process is a version of the Euclidean algorithm, which begins to explain the connection between the continued fraction of the divergence and the principal pairs of the lattice that we will see in subsequent chapters.

We gave a standard version of the Euclidean algorithm back in Sect. 3.4.1 as a series of successive reductions of pairs of integers, and it can be generalised to vectors as long as we choose a norm for those vectors. Another way of viewing the first half of this chapter and Theorem 4.3 in particular is that if we define the size of a lattice vector as its rise, then application of the Euclidean algorithm to successively reduce the size of pairs of generating vectors $\mathbf{p}_m$ and $\mathbf{p}_n$ terminates with a pair that have the very smallest rises, ie $\mathbf{p}_0$ and $\mathbf{p}_1$. If instead we define the size of a lattice vector as its length, then the same Euclidean algorithm terminates not at the vectors with shortest rise, but the vectors with shortest length: the principal vectors. One step of the algorithm is illustrated in Fig. 4.11.

Specifically, Turing-Euclid reduction starts with $i = 0$ and two vectors $\mathbf{a}$ and $\mathbf{b}$ that generate the lattice, and ordered so that $|a| > |b|$. Set $\mathbf{r}_{-1} = \mathbf{a}$ and $\mathbf{r}_0 = \mathbf{b}$ and then

1. Choose q_i to be the integer that minimises $|\mathbf{r}_{i-1} - q_i \mathbf{r}_i|$. If there is a tie, choose the q closest to zero.
2. If $q_i = 0$, then set $N = i$ and terminate.
3. Otherwise set $\mathbf{r}_{i+1} = \mathbf{r}_{i-1} - q_i \mathbf{r}_i$. Increment i and return to step 1.

As before $|\mathbf{r}_i|$ is strictly decreasing over a finite set of vectors shorter than $\mathbf{a}$ so the algorithm terminates with an $\mathbf{f}$ and $\mathbf{s}$ satisfying

$$|\mathbf{f}| \leq |\mathbf{s}| \leq |\mathbf{s} - q\mathbf{f}| \text{ for any integer } q. \tag{4.3}$$

4.16 In the lattice $h = 1/100, d = 17/72$, find an example where Turing-Euclidean reduction requires more than two steps to terminate.

4.12 Turing-Euclid Reduction of Generating Pairs

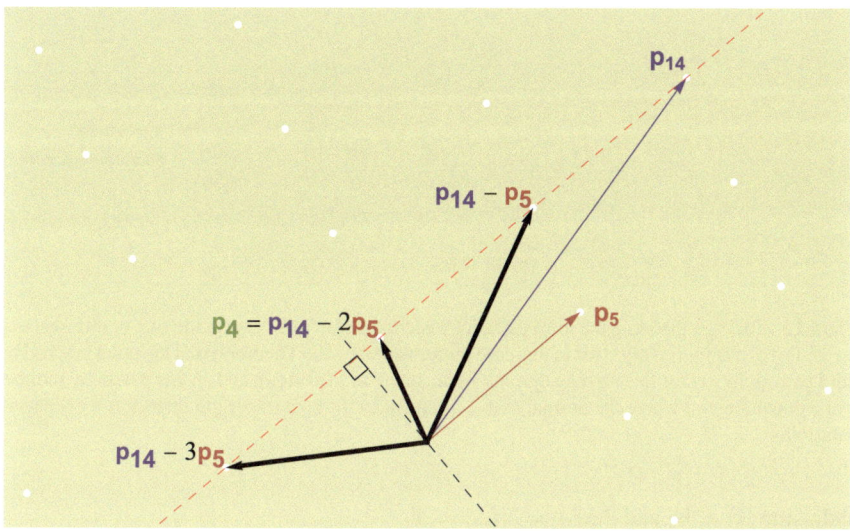

Fig. 4.11 A pair of generating vectors reduced by a step of the Euclidean algorithm. $\mathbf{p}_4 = \mathbf{p}_{14} - 2\mathbf{p}_5$ is the shortest of all vectors of the form $\mathbf{p}_{14} - q\mathbf{p}_5$ and by construction it must be shorter than $\mathbf{p}_5$; the pair (5,14) has been reduced to the pair (4,5)

Turing-Euclid reduction certainly terminates with a pair of vectors no longer than the pair that it started with. To show that it actually terminates at the shortest pair of vectors in the lattice we will use

Theorem 4.7 *If* $\mathbf{f}$ *and* $\mathbf{s}$ *are a pair of vectors with positive rise that generate the lattice and satisfy* (4.3) *then they are the first and second principal vectors and the third principal vector has the form* $\pm\mathbf{f} \pm \mathbf{s}$.

Proof First note that (4.3) with $q = 1$ shows $|\mathbf{s}|^2 < |\mathbf{s} - \mathbf{f}|^2 = |\mathbf{s}|^2 - 2|\mathbf{f}.\mathbf{s}| + |\mathbf{f}|^2$ and so $|\mathbf{f}.\mathbf{s}| < \frac{1}{2}|\mathbf{f}|^2$.

Set Δ to be the sign of $\mathbf{f}.\mathbf{s}$, and let $\mathbf{t} = \mathbf{s} - \Delta\mathbf{f}$. We will show that no other lattice vectors, not co-linear with $\mathbf{f}$ or $\mathbf{s}$, are shorter than $\mathbf{t}$. Since $\mathbf{f}$ and $\mathbf{s}$ are generating, we can write any other vector as $\mathbf{x}(q, r) = q\mathbf{s} - r\Delta\mathbf{f}$ for nonzero integer q, r, where we have used $\Delta = \pm 1$. We can show that $\mathbf{x}(q, -r)$ is longer than $\mathbf{x}(q, r)$ by using

$$|\mathbf{x}(q, -r)|^2 - |\mathbf{x}(q, r)|^2 = 4qr|\mathbf{f}.\mathbf{s}|, \qquad (4.4)$$

so we need only check $r > 0$. Setting $q = 1$ in Eq. 4.4 we have in particular that $|\mathbf{t}| = |\mathbf{s} - \Delta\mathbf{f}| < |\mathbf{s} + \Delta\mathbf{f}|$. We calculate

$$|\mathbf{x}(q, r)|^2 - |\mathbf{t}|^2 = (q^2 - 1)|\mathbf{s}|^2 - 2\Delta(qr - 1)(\mathbf{f}.\mathbf{s}) + (r^2 - 1)|\mathbf{f}|^2$$

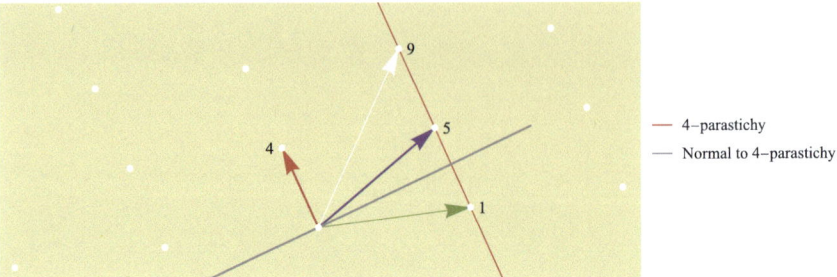

Fig. 4.12 Turing's theorem: the third parastichy vector in a lattice is always the sum or difference of the first two. It is intuitively clear in this case (and ensured by the Theorem) that the point reached by third parastichy vector lies on the adjacent 4-parastichy to the origin, and so has paastichy number 9 or 5 depending on where the normal to the 4-parastichy through the origin intersects the adjacent parastichy

and using $|\mathbf{f}| < |\mathbf{s}|$ and $2\Delta \mathbf{f}.\mathbf{s} = 2|\mathbf{f}.\mathbf{s}| < |\mathbf{f}|^2$

$$|\mathbf{x}(q,r)|^2 - |\mathbf{t}|^2 \geq \left(q^2 - 1 - (qr - 1) + r^2 - 1\right)|\mathbf{f}|^2$$
$$= (q^2 - qr + r^2 - 1)|\mathbf{f}|^2 = ((q-r)^2 + qr - 1)|\mathbf{f}|^2$$

and the final bracket is nonnegative since $q, r \geq 1$. This shows that $\mathbf{f}$ and $\mathbf{s}$ in order are the first two principal vectors and that the third is $\mathbf{s} - \mathbf{f}$ if $\mathbf{f}.\mathbf{s} > 0$ and $\mathbf{s} - \mathbf{f}$ otherwise. □

The algebra of the proof could be simplified by first rotating and scaling the lattice so that $\mathbf{f}$ is horizontal and of length 1: this is the renormalisation transform we will encounter in the next Chapter.

Because the first and second parastichy vectors $\mathbf{f}$ and $\mathbf{s}$ are of the forms $\mathbf{p}_m$ and $\mathbf{p}_n$, with parastichy numbers m and n, a simpler restatement of this result is Turing's theorem :

Theorem 4.8 *The third parastichy number is the sum or difference of the first two.*

See Fig. 4.12. This form of the Theorem neglects the hats, so requires us to interpret $1 + \hat{1}$ as 2 or $\hat{2}$ as necessary. We will see in the next Chapter how Turing sought to use this theorem as part of his explanation of Fibonacci structure.

4.13 Orthostichies

With the notation in place we can observe a consequence of having golden angle lattices. An *orthostichy* is a parastichy vector of nonnegative rise which is closer to the vertical than any other parastichy vector of lower nonnegative rise.[3] Thus

---

[3] Historically, orthostichy has also been used to mean a strictly vertical parastichy.

4.14 Touching-Circle and Hexagonal Lattices

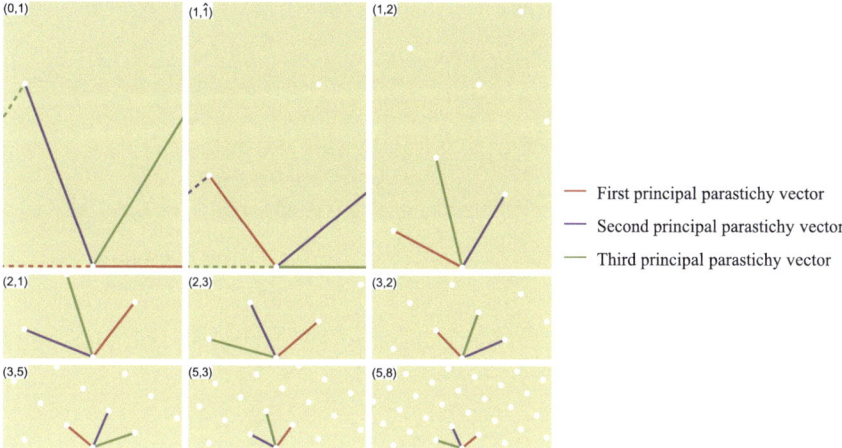

Fig. 4.13 Increasing Fibonacci pairs in a golden angle lattice as the rise is reduced

there are is a sequence of orthostichies, starting with $\mathbf{p}_0$ and $\mathbf{p}_1$, of increasing rise. If the divergence of the lattice is a rational $d = p/q$ in its lowest terms then the final orthostichy is $\mathbf{p}_q$; if d is irrational then the sequence of orthostichies is infinite, and tends to the vertical. The parastichy number of each orthostichy corresponds to the denominator of an increasingly accurate series of rational approximations to the divergence. Provided the rise is small enough (and excluding spiral lattices), the principal parastichy pair of the lattice will be two successive orthostichies, and as the rise continues to decrease the principal parastichy counts will move up the sequence of these denominators. Thus the principal parastichy numbers could be made to be successive and arbitrarily large Fibonacci numbers down to an arbitrary small h by setting the divergence sufficiently close to the golden angle, as in Fig. 4.13.

The Standard Picture models have the divergence as an outcome, not an input. So under the Standard Picture, this observation explains why, if we see some Fibonacci structure, we see a lot, not why we see it in the first place.

4.14 Touching-Circle and Hexagonal Lattices

Equipped with the ability to find the shortest vectors in a lattice we can now classify lattice types, as in Fig. 4.14.

Definition 4.6 A hexagonal lattice is a lattice where the first three parastichy vectors are all the same length.

4.17 Show that every hexagonal lattice is hexagonal.

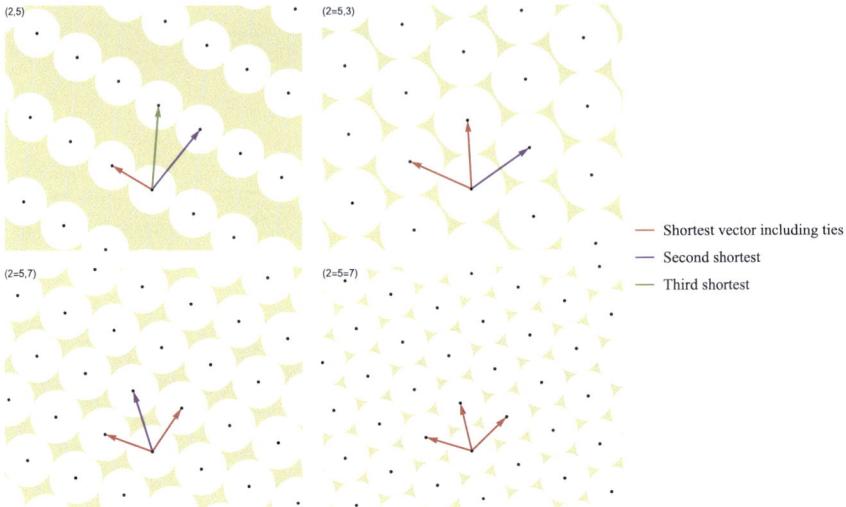

Fig. 4.14 Different types of lattice. (2,5): a lattice with principal vector $\mathbf{p}_2$ and second principal $\mathbf{p}_5$; (2=5,3): a touching-circle lattice with a tie for the two shortest vectors, which in this case are non-opposed; (2=5,7) a touching-circle lattice which is opposed; and (2=5=7), a hexagonal lattice

Less fundamental, but sometimes of use to identify, are square lattices:

Definition 4.7 An square lattice has the first and second parastichy vectors the same length and at right angles.

Square and hexagonal lattices are special cases of van Iterson lattices:

Definition 4.8 A van Iterson, or touching circle, lattice, is one in which the first and second parastichy vectors have the same length.

The van Iterson connection will appear in the next chapter; the reason for the name 'touching circle' is that disks of diameter equal to the first principal parastichy vector can be placed at each point of a van Iterson lattice without overlapping, and will touch at a point halfway along that vector.

Figure 4.15 shows a number of touching circle lattices, especially showing examples of square and hexagonal lattices.

In a square lattice the first two parastichy vectors $\mathbf{p}_m$ and $\mathbf{p}_n$ have the same length, and the third and fourth parastichies also share the length $\sqrt{2}|\mathbf{p}_m|$; this is the boundary case when the third parastichy number changes between the sum and the difference of the first two. In a hexagonal lattice, the first, second and third principal vectors all have the same length, and can be written as $\mathbf{p}_m$, $\mathbf{p}_n$, and $\mathbf{p}_{m+n}$ with $m < n$; this is the boundary case between changes of first and second principal parastichy numbers. By considering these boundary cases, we will be able to infer how principal parastichy numbers change as we cross the boundaries.

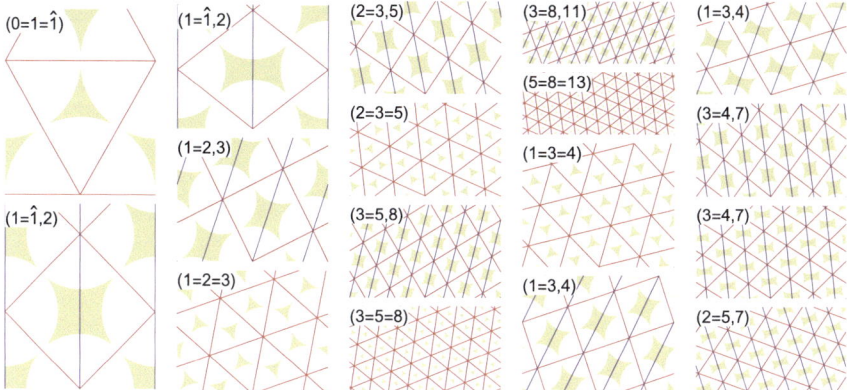

Fig. 4.15 Some touching circle lattices, labelled by the principal vectors. Red lines correspond to the shortest vectors in the lattice, including ties, and blue lines to one of the second-shortest vectors after ties. Redrawn from van Iterson [127]

4.15 Multijugate Lattices

The *jugacy* of a cylindrical lattice is the number of points in the lattice at each rise. The assumption at the beginning of this chapter that of nozero rise per node ensured that we have so far only seen monojugate lattices: those with a jugacy $J = 1$. More generally, a multijugate lattice $\mathcal{L}(d, h, J)$ can be defined as the set of points $(kd + l/J - [kd + l/J], kh)$ with $k \in \mathbb{Z}$ and $l \in 0, \ldots, J$, and can be constructed by placing J copies of the primary cylinder next to each other and then reducing the divergence by a factor of J so the multijugate lattice is still on a cylinder of circumference 1: see Fig. 4.16. If we then also reduce the rise by a factor of J we will produce a lattice in which the parastichy vectors have the same proportions as in the original. Instead of numbering the points 0, 1, ..., at each succesive rise, we can number them around the cylinder first and then increase with the rise. With this natural numbering of points in the multijugate lattice, the points numbered 0 and k in the monojugate lattice become the points 0 and kJ, so the principal parastichy numbers of a the lattice $\mathcal{L}(d/J, h/J, J)$ are J times the principal parastichy numbers of $\mathcal{L}(d, h, 1)$.

4.16 Dropping the Hats

Complementary vectors and the hat notation have done their job now in allowing precision in the Theorems of this Chapter when applied to spiral lattices and we can drop them from now on, with the understanding that any future description of a (m,m) parastichy pair is referring to what this chapter called a (m,m̂) pair. In

Fig. 4.16 Left: A monojugate lattice with divergence $d = 2/10$, rise $h = 1/10$, and $J = 1$, which has principal parastichy numbers $(1,4)$. Right a bijugate lattice with $J = 2$, divergence d/J, rise h/J, whose principal parastichy numbers are $(2,8)$

particular we can without ambiguity now talk of the lattice with divergence $d = 1/2$ as having a $(1,1)$ parastichy pair corresponding to the two lines from the origin to $1/2, h$ winding in opposite directions around the cylinder.

4.17 Relation to *Phyllotaxis: A Systemic Study*

Useful contributions to the analysis of cylindrical lattices then came from many writers, most relevantly a series of papers in the 1970s by Adler. Details of these and many further references can be conveniently found in Jean's 1994 book *Phyllotaxis: a systemic study in plant morphogenesis* [60]. This section notes a few of the differences between Jean's book and the current one and can be skipped by most readers. Jean's book included an invaluable mathematical synthesis, predominantly but not exclusively of Adler's work, but perhaps as a consequence contains a handful of mathematical inconsistencies. It is a tribute to the centrality of Jean's textbook to the field that it is still worth untangling them; and while in general the problems are ones that do not apply to large-Fibonacci-number lattices, the viewpoint of the next chapter requires a robust foundation for simpler lattices too.

Although I have preserved Jean's name for what is here Theorem 4.5 as the 'fundamental theorem of phyllotaxis', it is instead Turing's theorem, Theorem 4.8, which is more fundamental from the bifurcation theory viewpoint of this book. This is because Turing's theorem allows patterns to be interpreted as the outcome of a dynamic, and modellable, biological process, while the 'fundamental' theorem assumes the very existence of the emergent lattice which we are attempting to explain.

4.17 Relation to *Phyllotaxis: A Systemic Study*

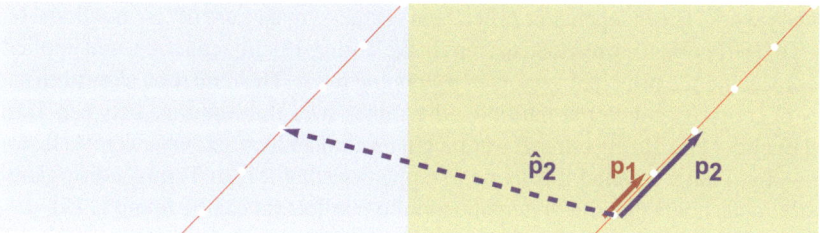

Fig. 4.17 A counter-example to Jean's Proposition A4.2 [60], with $m = 1, n = 2, d = 7/72$ and $h = 0.1$; this counter-example holds for all h and $0 < d < \frac{1}{4}$. A4.2 asserts that the 1-parastichy and the 2-parastichy form what Jean calls a visible pair and we call a generating pair. For spiral lattices, in the sense of Sect. 4.11, Jean's A4.2 fails because the 2-parastichy corresponding to $\mathbf{p}_2$ is not visible as it is parallel to $\mathbf{p}_1$; it is the pair $(1,\hat{2})$ comprising $\mathbf{p}_1$ and $\hat{\mathbf{p}}_2$ which is the generating pair

One notational difference from Jean is in the idea of the generating pair which Jean makes much of use of under the name of the 'visible pair'. Jean gives a number of useful theorems about generating pairs: his A4.1 on p. 306, for example, is in close correspondence with our Theorem 4.3. These theorems are correct and consistent for most phyllotactic lattices but in my interpretation they fail in some edge-cases alluded to in Exercise 4.13. Specifically, they fail with the spiral lattices we introduced in Sect. 4.11. As a counter-example to Jean's A4.2 (p. 307 of [60]), consider its statement for the particular case $m = 1$ and $n = 2$. It becomes 'the parastichy pair $(1,2)$ is visible and opposed iff there exist unique integers u and v such that $0 \leq u < 1, 0 \leq v < 2$, $|v - 2u| = 1|$ and such that the divergence $d < \frac{1}{2}$ lies at or is between u and $v/2$'. Solving the constraints unambiguously gives $u = 0$ and $v = 1$, and gives a generating interval of $[0, 1/2]$. However the pair $\mathbf{p}_1$ and $\mathbf{p}_2$ are only generating on the interval $d \in [1/4, 1/2]$. For any $d < 1/4$ it is not true, at least in the language of this book, to say as Jean's A4.2 does that the parastichy pair $(1,2)$ is a generating pair (and also incorrectly, that the pair is opposed for all $0 < d < \frac{1}{2}$). What has happened is that Adler's proof method actually leads to the pair $(1,\hat{2})$, ie the 1-parastichy vector and the 2-complementary vector, which *is* generating and opposed: see Fig. 4.17.

Another area where this text differs from Jean's synthesis is the *conspicuous pair* as a model of the most obvious spirals. Adler [1] defined a conspicuous visible opposed parastichy pair as identical to our principal parastichy pair (at least when opposed, and away from spiral lattices), and in 1988 [59, p. 214] Jean followed this, but by 1994 he was differently defining the conspicuous pair as that closest to a right angle [60, p. 18].

In any case I prefer here to use the principal parastichy pair as the most natural model of 'obvious'. The concept of the principal parastichy pair, central to this book, was first made explicit by Turing [124] in unpublished material in the early 1950s: Theorems 4.7 and 4.8 are based on his ideas. Turing's algorithm was eventually published in the 1990s as described in [124]. The theorem I have called here Turing's Theorem is, with hindsight, implicit in or a simple consequence of Bravais' nineteenth century work. The independent Adler/Jean synthesis makes use of 'contractions' (e.g. A4.3 of [60]) which correspond to steps in the Turing-Euclid reduction, and implicitly uses reaching the principal vectors as a stopping point. The reduction algorithm itself was of course encountered multiple other times over the centuries between Turing and Euclid—and Turing himself was taught by number theorists who were well aware of it—but I have adopted this name to emphasise that it is in Turing's disorganised posthumous notes that the first application to phyllotaxis can be found [115].

Adler was well aware of Turing's notes on phyllotaxis and cited them in his 1974 literature survey [1]. Adler's perspective was that the central mathematical problem was the relationship between the divergence and the set of possible generating pairs. It was from this perspective that he wrote that Turing had been 'prevented' from using this 'fundamental' concept because of Turing's focus on the principal parastichy pair, which is in fact more fundamental to the modern Standard Picture. For some reason, perhaps the very long delay in publishing Turing's notes, even this dismissal was removed from Jean's book, even though the notes had by then been published. As a consequence Jean's book contains no explicit equivalent of Turing's Theorem.

4.18 Summary

We have classified parastichy pairs in lattices. Most simply we classified them by whether they are opposed, that is wind in opposite directions. That simple classification is less useful than knowing whether they are generating, that is capable of describing the whole lattice, and we saw that a simple visual check for this was whether the triangle they defined contained any other lattice points. Each lattice has many generating pairs but typically only one of them is the shortest pair.

We called this shortest pair the principal parastichy pair and argued that this a good model for the spiral counts typically done when examining specimens. This principal parastichy pair depends on the rise: lattices with the same d and different h will have exactly the same generating pairs but which of those is the principal pair will vary with h. Crucially, we observed Turing's Theorem: geometrical constraints mean that the third principal parastichy number is always the sum or difference of the first two.

Chapter 5
Classifying Cylindrical Lattices

Abstract The previous chapter made precise the ideas of a lattice $\mathcal{L}(d, h)$, where we take the divergence d and the rise h as parameters, and the principal parastichy pair as the pair of integers corresponding to the two shortest informative vectors in the lattice. In this chapter we solve the problem of classifying every lattice by its principal parastichy pair. The answer is given in the van Iterson diagram of Fig. 5.1 and the rest of the chapter explains how this Figure is constructed and its implications. Since the classification changes at the points of lattice space where either the first and second, or the second and third, parastichy vectors are equal in length, we need to find the d and h at which this occurs. This requires us to identify *touching-circle lattices*, and we will first construct Fig. 5.1 by algebraically finding touching-circle lattices, and then see how a more powerful and more abstract approach based on lattice re-normalisation gives the same answer and explains much of the geometric structure of the diagram.

5.1 Components of the Van Iterson Diagram: The m=n Branch

We have not yet seen shown how to draw the van Iterson diagram nor seen a reason for its remarkable self-similarity. We will do the latter in Sect. 5.4, and though elegant and powerful, that approach masks the structure of the crucial lattice bifurcation. So first we take an approach using only elementary algebra, and then only afterwards see why the answers take the form they do. Figure 5.2 redraws the van Iterson diagram in a subset of lattice space, and adds labels for the boundaries and the bifurcation points at which they meet (Fig. 5.1). It shows that we can construct the structure of lattice space by finding the arcs on which the principal vectors have the same length. We call these branches:

Definition 5.1 The branch m=n is the set of solutions in lattice space on which $\mathbf{p}_m$ and $\mathbf{p}_n$ are a generating pair and have the same length.

Recalling from 4.14 that a van Iterson or touching circle lattice has the first two parastichy vectors the same length, an m=n van Iterson lattice lies on a m=n

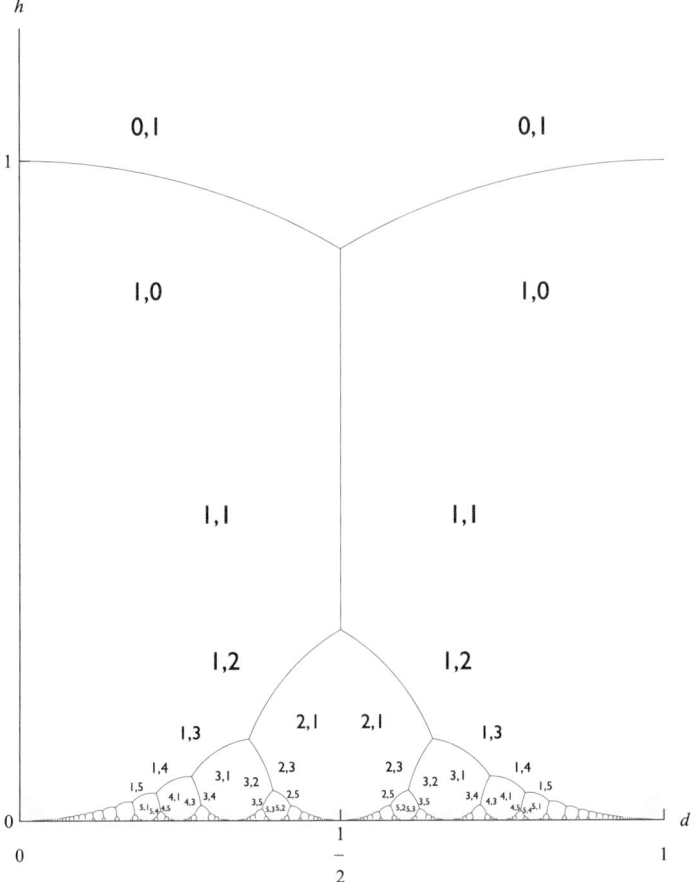

Fig. 5.1 Lattice space classified by principal parastichy numbers (m, n). The previous chapter's distinction between parastichy vectors and complementary vectors is dropped so that eg the (1,2) region includes lattices where the principal parastichy pair is (1, 2) but also lattices where it is (1,$\hat{2}$)

branch, but not all lattices on the branch are van Iterson. Typically the branches are semicircular and the van Iterson region, where the two generating vectors are also the two shortest vectors, is a single segment on the circle.

This section, then, finds the d-h relationship on the m=n branch. If a pair of parastichy vectors are equal in magnitude, $|\mathbf{p}_m| = |\mathbf{p}_n| = 2r$, then

$$4r^2 = x_m^2 + (mh)^2, \tag{5.1}$$

$$4r^2 = x_n^2 + (nh)^2. \tag{5.2}$$

In addition, for each of $\Delta = \pm 1$ we can find winding numbers (u, v) such that $vm - un = \Delta$ and then

5.1 Components of the Van Iterson Diagram: The m=n Branch

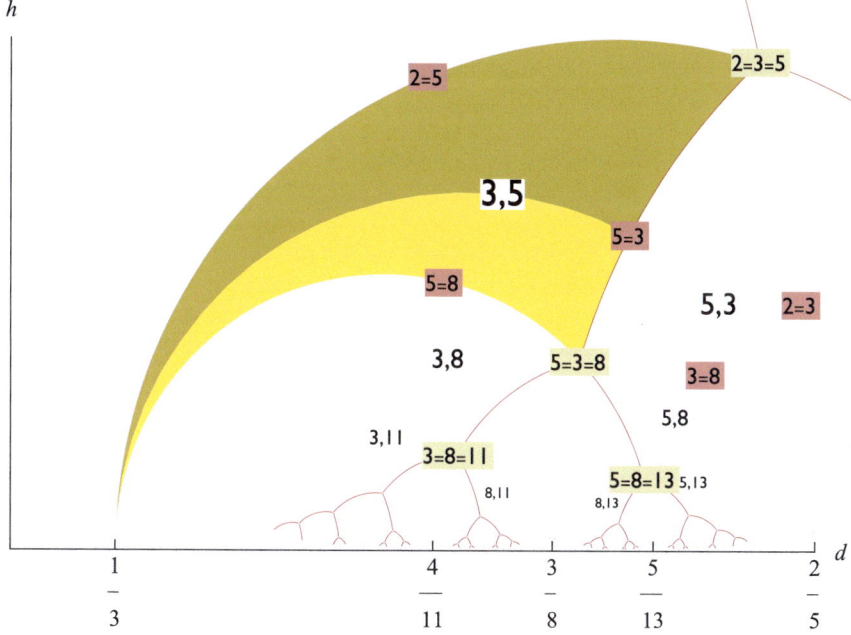

Fig. 5.2 Lattice space is partitioned into regions with the same principal parastichy pair by arcs on which two parastichy vectors have the same magnitude. These m=n branches are semicircles; the branches meet when the lattice is hexagonal, labelled in pink. A red line shows when the m=n branch is a van Iterson or touching circle lattice. Regions are labelled by their principal parastichy numbers m, n; the topological properties of the Figure do not rely on the numbers being Fibonacci. d-co-ordinates of the endpoints of the branches are calculated here in the case $m = 2, n = 3, u = 1, v = 1$

$$x_m = md - u, \qquad (5.3)$$
$$x_n = nd - v. \qquad (5.4)$$

5.1 Find r^2 as a function of m, n, x_n, x_m and Δ.

With a little work we can find

$$h^2 + (d - \bar{d})^2 = (n^2 - m^2)^{-2}, \qquad (5.5)$$
$$4(n^2 - m^2)r^2 = 2n\Delta(md - u) - 1. \qquad (5.6)$$

where we have set $\bar{d} = (nv - mu)/(n^2 - m^2)$. The first of these shows that d and $h > 0$ lie on a semi-circle, and the second that r^2 changes linearly with d on that semi-circle, and thus is a decreasing function of h.

5.2 Show that (5.5) and (5.6) are correct.

5.3 Find the values of x_m, x_n, and r at the left and right intersections of the m=n branch with $h = 0$.

Since there are two different (u, v)'s depending on $\Delta = \pm 1$ so there are two different semicircles in (d, h) space linked by the $(d, \Delta) \to (1 - d, -\Delta)$ symmetry.

By construction the pair is generating, so the projection of the semicircle onto the d axis is a subset of the generating interval for (m, n). From the last Chapter, there can be at most one point on in this interval where the pair changes from opposed to non-opposed. This happens only when $x_m = 0$ or $x_n = 0$ but $x_m = 0$ means $x_n^2 = (m^2 - n^2)h^2$ so cannot occur if $m < n$ and we are left with $x_n = 0$ which implies that $d = v/n$.

5.2 Triple-Points

Definition 5.2 A triple-point (m=n=t) is a point in $\mathcal{L}(d, h)$ space at which the lattice is hexagonal, or equivalently where the lattice has the first three principal vectors equal in length, with three distinct corresponding parastichy numbers m, n, t.

The crucial organising centres of lattice space are the triple-points, where two of the equal-length branches m=n and n=t collide, and there is also a branch m=t through the point where by Turing's theorem t must be the sum or difference of m and n. There are two important special cases (0=1=$\hat{1}$) and (1=$\hat{1}$=2) and otherwise the triple-point is of the form (m=n=n+m) with $0 \leq m < n$. The semicircles defined in the previous section provide necessary but not sufficient conditions for a lattice to be a van Iterson lattice with $|\mathbf{p}_m| = |\mathbf{p}_n|$: we also need that $\mathbf{p}_m$ and $\mathbf{p}_n$ are the principal parastichy vectors. The diagram of Fig. 5.2 is organised by these triple-points at which the lattice has hexagonal symmetry as in e.g. the (2=3=5) lattice in Fig. 4.15. The bifurcations at these triple-points can completely characterise transitions in lattice space.

It is possible, as in Exercise 5.4, to calculate the positions of the triple-points algebraically, though this does not offer much insight. One relation that is worth noting is about the size of the disks that fit into the hexagonal lattice: these have radius r where

$$h = 2\sqrt{3}r^2 \tag{5.7}$$

5.4 Find a relation between r, m and n at a hexagonal lattice, and use this to show Eq. (5.7).

5.2.1 The m=n Branch in Lattice Space

Now we can describe the general structure of the m=n branch and its bifurcations, as shown in Fig. 5.2.

Theorem 5.1 *For co-prime $1 \leq m < n$, (but excluding the 1=2 branch) the m=n branch has exactly two triple points on it, one which is (m=n−m=n) and one which is (m=n=n+m), and the lattice is van Iterson only on the branch between these two points.*

Proof Theorem 4.7 guarantees that the third of the principal parastichy vectors must be either $\mathbf{p}_n - \mathbf{p}_m$ or $\mathbf{p}_n + \mathbf{p}_m$, and so be at the intersection of the branch with either the branch m=n−m or the branch m=n+m, but each of these branches are semicircles with a single intersection with the m=n branch, so there are exactly the two triple points. The m=n branch is not van Iterson for h small enough, so the van Iterson sub-branch must be between the triple points. □

It is simple to show from Exercise 5.4 that on the m=n branch with $m < n$, the triple point (m=n=m+n) always occurs at a lower value of h than the triple point (m=n−m=n).

Between each triple-point, there is a point on the branch where the lattice is square. As lattice parameters pass through these square points, the third and fourth parastichy numbers swap and to that extent these points also organise the lattice space. Square lattices also have a role in modelling obvious parastichies as some authors have considered the 'most obvious' to be those which are orthogonal.

5.5 At the top of the m=n branch, the three principal parastichy numbers change from (m=n,n−m) to (m=n,n+m) at the point where the lattice is a square lattice with principal parastichy numbers (m=n,n−m=n+m). Find the values of h, r and d at which this happens and hence show this is unique on the van Iterson branch.

5.3 Nearly Hexagonal Lattices: Unfolding the Triple-Point Bifurcation

We have established algebraically the structural properties of lattice space so that we can draw it, but the key property of lattice space we need to understand for Fibonacci structure depends only on what happens at triple-points at which the lattice is hexagonal. In fact we can obtain the key structural information by understanding the structure of lattice space near each such bifurcation: in the language of bifurcation theory we need only consider the unfolding of the triple-point.

In particular, we will show that one of the branches downward from the triple-point corresponds to a non-opposed principal parastichy pair. It is crucial to note that van Iterson lattices can sometimes have non-opposed principal parastichy vectors as shown in Fig. 5.3.

Fig. 5.3 A van Iterson lattice can be non-opposed. The two shortest vectors $\mathbf{p}_2$ and $\mathbf{p}_5$ with $|\mathbf{p}_2| = |\mathbf{p}_5|$, both wind in the same direction. This example has $d = 1/21(8 + \cos\theta)$ and $h = 1/21\sin\theta$ where $2\theta = \arctan(21\sqrt{3}/11) + \arctan(\sqrt{21}/2)$ and has been constructed by the renormalisation technique of Sect. 5.4

To classify the branches as they leave the triple-point, we need to know about small changes in the lattice away from that triple-point. That can be done algebraically by using the results of Sect. 5.1 or graphically by consideration of Fig. 5.4. Either of these show the following properties of a triple-point:

Theorem 5.2 *Suppose m, n are co-prime integers with $0 < m < n$ but not $m = 1, n = 2$. Then each of the two symmetric semicircles making up the branch m=n have two triple-points. At the triple-point (m=n=n+m), which is the lower of the two:*

1. *Close to, but above the triple-point, the branch m=n is van Iterson and opposed, m=n+m is non van Iterson and nonopposed, and n=n+m is non van Iterson and opposed.*
2. *Close to, but below the triple-point, the branch m=n is non van Iterson and opposed, m=n+m is van Iterson and nonopposed, and n=n+m is van Iterson and opposed.*

This triple-point is also the higher of the two triple-points for the nonopposed branch, so this branch will remain van Iterson until it passes through its lower triple-point. But we have just seen that it must be opposed there, and we know from the previous chapter that there is only one change of opposition in the d-generating interval, which the branch is contained in, so there is a unique point on the $m = n + m$ van Iterson sub-branch at which it ceases to be opposed. This point was exactly $d = v/n$.

5.3 Nearly Hexagonal Lattices: Unfolding the Triple-Point Bifurcation

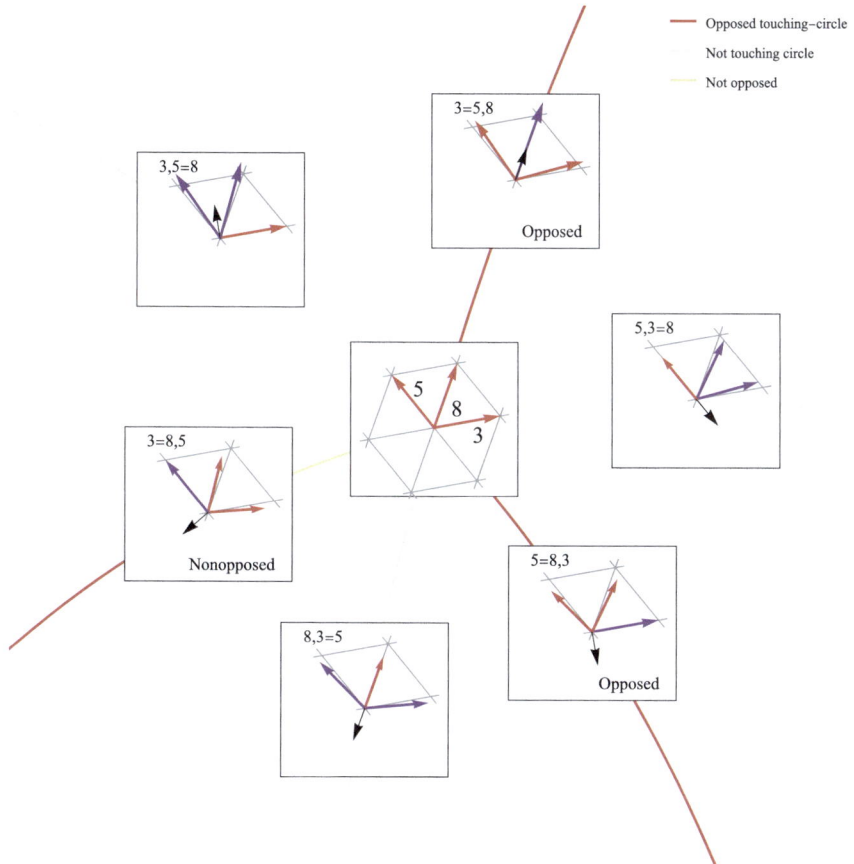

Fig. 5.4 Lattice deformations near a triple-point classified by Theorem 5.2. m=n branches are drawn as in Fig. 5.2. As in that Figure, the unfolding is true for general m, n but is labelled for clarity with $m = 3$ and $n = 5$. Red arrows: primary parastichies; blue arrows secondary parastichies. Grey arrows show the direction in which the hexagonal lattice is stretched so as to preserve two vectors of equal length. In particular note that the m,n+m van Iterson branch below the triple point is nonopposed but the n,n+m branch is opposed

These properties taken together prove that the structure visible in the numerically calculated Figures of this chapter do indeed hold for all lattices. We can also redraw in Fig. 5.5 the van Iterson tree, showing which portions are opposed and which nonopposed. This Figure illustrates an important consequence of Theorem 5.2: any model whose solutions move through the van Iterson tree but always reject non-opposed solutions in favour of opposed ones will produce Fibonacci structure.

Yet more algebra can be deployed to show that no triple-point occurs at the maximum of a branch, so all the van Iterson sub branches can be traversed with h strictly decreasing.

5.6 Show this. (This is easier to do after Sect. 5.4)

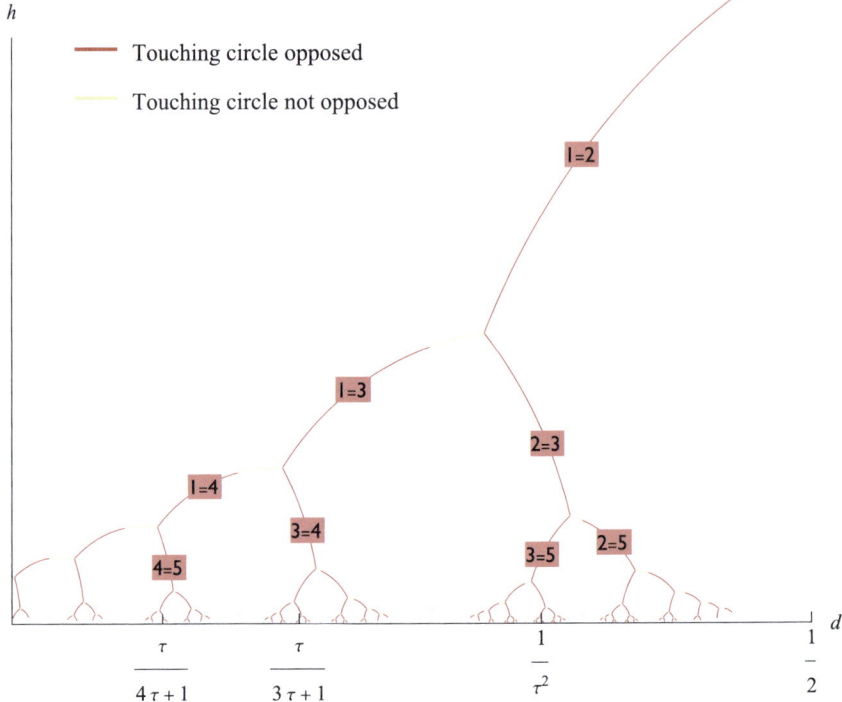

Fig. 5.5 At every bifurcation of the van Iterson tree, one downward branch is opposed and one is non opposed. Opposed branches are shown as red lines and non-opposed ones as yellow

We have shown that packing efficiency remains close to the hexagonal optimum on the van Iterson branch, but we also need the following:

Theorem 5.3 *Near but below the triple-point, the packing efficiency of the opposed branch, as a function of r, is higher on the opposed van Iterson than on the nonopposed van Iterson branch.*

Proof Take the triple-point (m=n=n+m) with $m < n$. For each of the branches, there is a relationship between the rise and the radius of the touching circle given by Sect. 5.2. On the opposed downward branch n=n+m this is $h_O(r) = h(r; n, n+m)$ and on the nonopposed m=n+m it is $h_N(r) = h(r; m, n+m)$. By setting the packing efficiency $P_O = \pi r^2/h_O$ and $P_N = \pi r^2/h_N$ on each branch, differentiating the difference with respect to r^2 and substituting back the value of r^2 at the triple-point we find

$$\frac{d}{dr^2}(P_O - P_N) = -\frac{8\pi}{9}\frac{(n-m)(m^2+mn+n^2)^3}{mn(m+n)} \quad (5.8)$$

Thus as r decreases below the triple-point, while the packing efficiency on the opposed branch also decreases, it does so more slowly than that on the nonopposed branch. □

5.4 Lattice Renormalisation

The algebra of the Chapter so far is straightforward, if a little tedious, but it doesn't give any insight into the remarkable self-similarity of the van Iterson diagram. The technique of lattice renormalisation uses the observation that one lattice can be mapped to another by scaling and rotating to give a powerful way of mapping from one part of lattice space to another. Lattice renormalisation gives considerably more insight into the structure of the van Iterson diagram than the results so far, at the cost of introducing the technology of Möbius functions, and although in my view it adds little to a scientific understanding of the appearance of Fibonacci numbers it does provide an description of lattice space.

We saw in the previous chapter that every cylindrical lattice unrolls to a plane lattice, but not all plane lattices have the right periodicity to be mapped down onto our cylinder of circumference 1. Specifically, a plane lattice can be collapsed to a cylindrical lattice using $(x, z) \to (x - [x], z)$ exactly when it contains the vector $(1, 0)$; and restricting to only monojugate lattices as we do adds the requirement that that must be the shortest vector in that direction; we called this being 1-periodic. Now suppose we have a cylindrical lattice with principal parastichy numbers (m,n). We can subject the corresponding plane lattice to a rotation, but in general the rotated lattice will not be 1-periodic. Suppose though that we choose the rotation that maps the principal parastichy vector $\mathbf{p}_m$ down onto the x-axis, followed by a uniform scaling that maps it onto the point $(1, 0)$: then by construction the transformed plane lattice is 1-periodic and there is a corresponding cylindrical lattice. At this point it's useful to represent lattice parastichy vectors by complex numbers $z_k = (kd - [kd]) + ikh$, because the transformation we have just described is exactly division of every lattice point by z_m; z here is a complex coordinate and not the real vertical coordinate of previous chapters. An example is given in Fig. 5.6. By a weak analogy with theoretical physics this rotate-and-scale process is called lattice renormalisation, and we call the corresponding parastichy vectors of the renormalised lattice w_m.

The original and the renormalised lattice are represented by two different points (d, h) in lattice space, and each point in that space has co-ordinates which are those of the parastichy vector of smallest positive rise of the corresponding lattice. So renormalisation corresponds to a map $w_1 = f(z_1)$; we will compute it, or rather its inverse g such that $g(w_1) = z_1$. This gives a way of constructing a lattice with any desired principal parastichy numbers m and n by multiplying a normalised one by a particular complex number z_m. Moreover, since the angle between the parastichy vectors is not changed by the renormalisation, a lattice with any desired such angle, such as a hexagonal lattice, can be found by starting from an lattice with the same angle in (0,1) space.

The simplest renormalisation maps a (0,1) lattice to (1,0) one through the transformation function $g(w_1) = 1/w_1$; and any lattice is invariant under the translation by an integer $g(w) = w + b$. So it is not a surprise that the function that achieves the renormalisation in general is a composition of these two function types and thus a Möbius map:

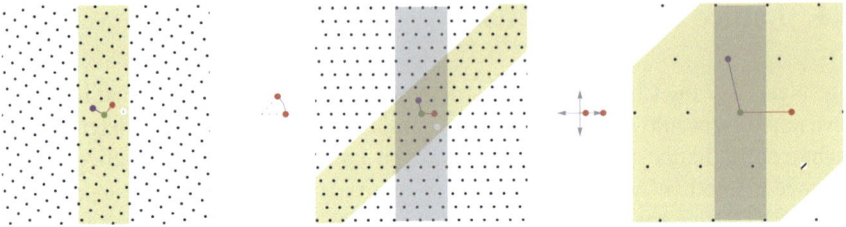

Fig. 5.6 A lattice with arbitrary parastichy numbers can be constructed from a $(0, 1)$ lattice by a carefully chosen scaling and rotation that preserves the cylindrical periodicity. From left to right, a lattice (here with principal parastichy numbers 3 and 2) is rotated so that the principal parastichy vector $\mathbf{p}_3$ becomes horizontal, then scaled to length 1 so it is the periodicity vector of a new lattice on the gray cylinder. The second principal parastichy vector, here 2, must map to the second shortest vector in the new lattice, which has principal parastichy numbers $(0, 1)$. Theorem 5.4 inverts this compound of rotation and scaling to compute the a complex number z_m which can multiply the $(0, 1)$ gray lattice and retrieve the desired (m, n) lattice. By construction, both lattices have the same angle between the principal parastichy vectors

Theorem 5.4 *Fix a pair of co-prime integers* (m, n), *and find a winding-number pair* (u, v) *with* $\Delta_{mn} = mv - nu$ *and* $|\Delta_{mn}| = 1$. *Label lattices by complex numbers so that* $\mathcal{L}(d + ih) = \mathcal{L}(d, h)$ *and for complex w define the Möbius map*

$$g_{mn}(w) = \frac{uw - v}{mw - n}. \tag{5.9}$$

Then if $\mathcal{L}(w_1)$ *is a* **(0,1)** *lattice and* $z = \Delta_{mn} g_{mn}(w_1)$, *then* $\mathcal{L}(z)$ *is an* **(m,n)** *lattice.*

Proof Writing vectors as complex numbers, take any w_1 in the **(0,1)** region of the van Iterson diagram, so $\mathcal{L}(w_1)$ has 1 as its shortest and w_1 as its second shortest vectors. We are going to rotate and scale this lattice by some z_m we need to find. The new lattice's shortest vector is the image of 1, i.e. z_m. The second shortest vector is the image of w_1 and is called $z_n = z_m w_1$. The new lattice is also generated by 1 and some vector we will call z_1; we don't yet know which vectors in the original lattice these are the images of under the rotation. For the new lattice to have principal parastichy numbers m and n we need to choose z_m so that z_m, z_n and z_1 satisfy

$$\begin{pmatrix} z_n \\ z_m \end{pmatrix} = \begin{pmatrix} n & -v \\ m & -u \end{pmatrix} \begin{pmatrix} z_1 \\ z_0 = 1 + 0i \end{pmatrix} \tag{5.10}$$

We invert this matrix equation, using the fact that winding-number pairs satisfy the Bézout relation with $\Delta_{mn} = mv - nu$ and $|\Delta_{mn}| = 1$, to get

$$\begin{pmatrix} z_1 \\ 1 \end{pmatrix} = -\Delta_{mn} \begin{pmatrix} u & -v \\ m & -n \end{pmatrix} \begin{pmatrix} z_n \\ z_m \end{pmatrix}, \tag{5.11}$$

and the ratio of these two rows gives the Möbius map g of the theorem:

5.4 Lattice Renormalisation

$$\frac{z_1}{1} = \frac{u z_n/z_m - v}{m z_n/z_m - n} \tag{5.12}$$

$$= g_{mn}\left(\frac{z_n}{z_m}\right) \tag{5.13}$$

But we have set $z_n = w_1 z_m$, so we now find $z_1 = g_{mn}(w_1)$ and can use (5.10) to find $z_m = m g_{mn}(w_1) - u$. This z_m defines the rotation and scaling we need so $\mathcal{L}(z_1)$ has shortest vectors $z_m = m z_1 - u$ and $z_n = n z_1 - v$ and is an (m,n) lattice.

So we could use z_1 as the z of the theorem. But as we show below, the imaginary part of z_1 has the sign of Δ_{mn}, and would not necessarily be a vector of positive rise. Since $\pm z_1$ each generate lattices with the same required parastichy numbers and related by reflection in $d = 0$, we set $z = \Delta_{mn} z_1$ to maintain the convention that z has positive rise. □

The Möbius map g may include an odd or even number of inversions in the unit circle, and it maps the upper half complex plane to either the upper or lower half depending on the sign of Δ_{mn}.

5.7 Suppose that $w = d_0 + i h_0$ with $h \geq 0$. Compute the real and imaginary parts of $z = \Delta_{mn} g_{mn}(w)$ and show that the imaginary part of z is always non-negative.

So for co-prime m and n we can construct a rotated lattice $z_m \mathcal{L}(w_1)$ with shortest vectors z_m and z_n. It is possible that these shortest vectors are not parastichy vectors in the sense specific to the previous chapter, but are instead complementary vectors. But the hatless parastichy counts will still be (m, n). In general, any u, v which satisfy the Bézout relation will work to construct a region of lattice space with the correct parastichy numbers although if the (u, v) are not winding numbers that region is an integer horizontal translation of the basic (m,n) region.

We can use the transformation g_{mn} to calculate the regions of lattice space as shown in more detail in Fig. 5.7. In general there are two choices for the winding-number pair for (m, n) and these correspond to the two (m,n) regions linked by the reflection symmetry.

Once we have done all this work to see where the renormalisation transform comes from, there is a payoff in simple characterisations of the van Iterson diagram as illustrated by the following exercises. Each of these results were originally found without the renormalisation transform, but now the mathematical structure becomes clearer.

5.8 Suppose that m, n, u, v are integers satisfying $mv - nu = \Delta$ with $|\Delta| = 1$. Find (m,n) lattices which are orthogonal and show they lie on circles in van Iterson space.

5.9 Find (m,n) lattices which are touching-circle and the special touching-circle cases of square and hexagonal lattices.

5.10 Find the magnitude of the principal parastichy vector in an (m,n) lattice with angle θ between its principal parastichy vectors. Find the radius of the disks in a touching circle lattice as a function of θ, and the divergence d as a function of r.

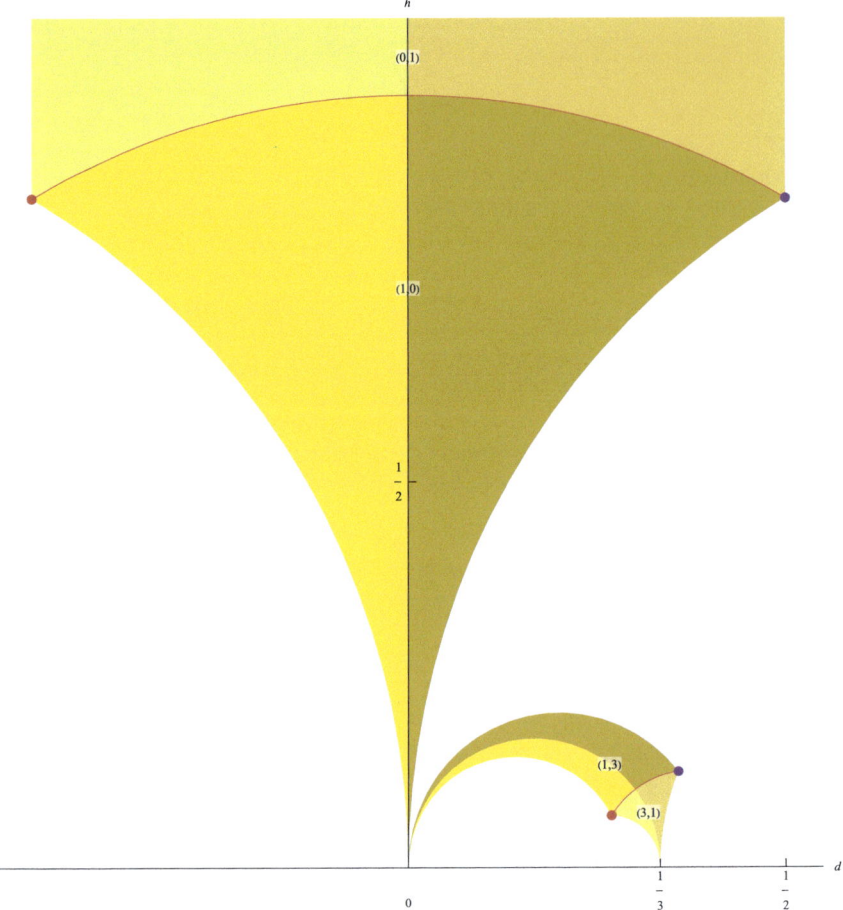

Fig. 5.7 The (1,3) region of lattice space is the image of the (1,3) region under the g_{mn} of Eq. (5.9) for $(m, n) = (1, 3)$

5.11 Suppose that the Fibonacci branch of the van Iterson tree is followed as r is decreased. Show that the larger parastichy number becomes F_{k+1} at approximately $r^{-1} = \sqrt{2} F_k$.

5.12 Compute the centre and the radius of the circle in van Iterson space on which the $m = n$ branch lies.

5.13 If m and n are co-prime and $m < n$, find the region where a (m,n) touching-circle lattice is non-opposed.

5.5 Classification of Van Iterson Space by Euclidean Reduction

It is no accident that there is a direct similarity between the coefficient matrix of the g_{mn} and that of the solution matrix for the Euclidean algorithm of Sect. 3.3. Each transition from a region of the van Iterson diagram with principal pairs (m,n) to the region above with pairs (m,n−m) is a substep of the Euclidean algorithm; this moving upwards by subtraction from the larger can be done exactly q_0 times to get $n' = n − q_0 m$ and reach the region (m,n') before we have to swap and repeat. Each (m,n) region of van Iterson space corresponds to a particular sequence of $q_1, q_2, \ldots q_N$s that take the region back to the (0,1) region. Moving down the tree through the $(r_i, r_i − 1)$ principal parastichy pairs, the Euclidean algorithm for u_i and v_i also yields $r_{i−1} v_i − r_i u_i = 1$ for each pair, and allows us to calculate the Fundamental Theorem of Phyllotaxis with ease. The Fibonacci branch, where we swap over at each transition, corresponds to all of the integer pairs whose q-sequence is exactly 1, which are the convergents to the golden Ratio.

One definite virtue of this analysis is to explain why the sign parameter Δ persistently emerges from the algebra of this Chapter. It shows that $\Delta_{mn} = (-1)^{(N-1)}$ where N is the number of steps in the Euclidean reduction of m and n.

It is a matter of taste whether the structure of van Iterson space as revealed by the modular group of transformations g_{mn} can be said to explain the prevalence of Fibonacci numbers: to my mind it is an attractive fact that Fibonacci lattices are in a sense the most invariant under these transformations, but I have not found any way to make use of the transformations—beyond code to draw the diagrams in this book—in connecting the models to biological data.

5.6 Packing Efficiencies

Some of the less convincing arguments for the occurence of Fibonacci phyllotaxis have centred on analyses of packing efficiency, partly because of phenomenon like the ones illustrated in Fig. 5.8. Each of the three node patterns in Fig. 5.8 contain the same number of nodes in the same sized region: in that sense mean packing is equivalent between them. What (extremely close) adherence to the golden angle offers is a closeness to uniformity in the angle distribution of the nodes. This most-uniform packing property, which is intimately related to the continued fraction representation of the golden angle, has been known for a long time and seems to have first been explicitly proved by Wright [132]. It has sometimes been suggested that this kind of uniformity in the distribution of leaf positions offers a fitness advantage in that such a leaf architecture could maximise sunlight capture. There's never been any good evidence of this inherently implausible fitness effect.

Nevertheless, given that there clearly are molecular mechanisms that promote particular kinds of close packing in node placement, it is worth clarifying how packing

Fig. 5.8 A golden angle lattice packs more uniformly than nearby lattices, in the sense that the angular distribution of the angles of the nodes is most even at the golden angle. Above, nodes with polar coordinates $(r, \theta) = (nh, n\phi)$ are shown for $n = 1 \ldots 1000$, $h = 0.001$, and $\theta = \tau/1.001$, τ and 1.001τ. Below, the angles of each node are binned and coloured by count

closeness varies over the van Iterson diagram. If we draw a circle of the same radius around every point of a lattice and choose the radius as large as possible without two circles intersecting, then the radius of such a circle is $|\mathbf{p}_m|/2$ where m is the primary parastichy number. There is one such circle of area $\pi|\mathbf{p}_m|^2/4$ attached to each point, but we saw in the previous Chapter that the cylinder area per point is h so the packing efficiency β of the lattice, defined as the fraction of the cylinder covered by the circles, is

$$\beta = \frac{\pi|\mathbf{p}_m|^2}{4h}$$

For the hexagonal lattice, the previous result that $h = 2\sqrt{3}r^2$ recovers the standard result that the packing efficiency of the hexagonal lattice is $\pi/2\sqrt{3}$; this implies that on every path through van Iterson lattices the packing efficiency keeps increasing and then decreasing so as to pass through this value at every triple-point.

5.14 Show that the packing efficiency of a van Iterson lattice is locally maximised at every triple-point.

We can calculate the packing efficiency β across van Iterson space by using the classification of this chapter to find $\mathbf{p}_m$. Within the $(0, 1)$ region, the primary parastichy vector is $(1, 0)$ of length 1, and β is simply $\pi/4h$. Since β is not changed by a rigid rotation and scaling of a lattice, we could also simply map this function $\pi/4h$ using the Möbius transforms into each region. Figure 5.9 shows a contour plot of the packing efficiency β along with a graph of β along a horizontal transect of lattice space. Since packing efficiency scales like $1/h$ in the $(0, 1)$ region, its contour lines there are horizontal, and so the contour lines of packing efficiency in each other, Möbius transformed region, become arcs of circles.

5.6 Packing Efficiencies

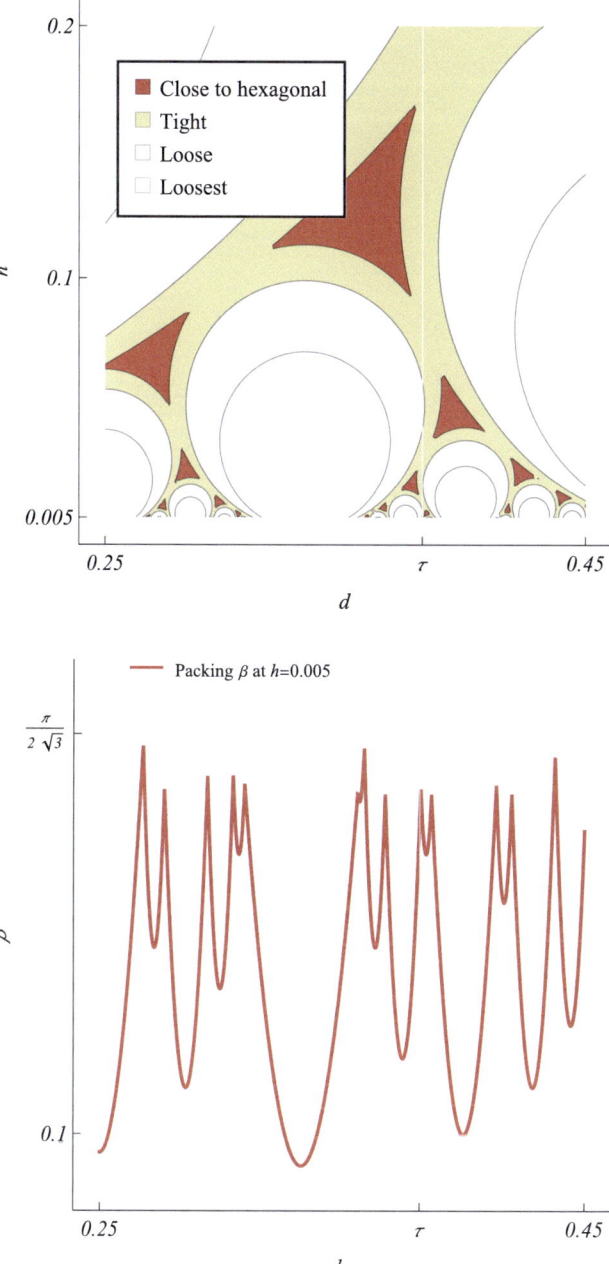

Fig. 5.9 The golden angle cannot achieve for maximal packing. Top packing efficiency β across van Iterson space (Close to hexagonal: $\beta > 0.8$, Tight: $\beta > 0.7$, Loose: $\beta > 0.6$, Looser $\beta < 0.6$). Bottom graph: β as a function of d across the line $h = 0.005$. Note that while the divergence $\tau/2\pi$ corresponding to the golden angle is a local maximum of β, it is not a global maximum

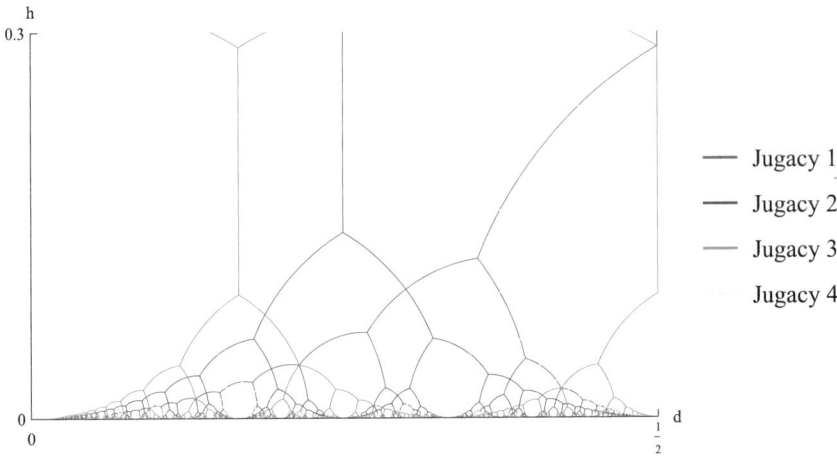

Fig. 5.10 van-Iterson trees for lattices of jugacies 1 to 4

5.7 The Structure of Multijugate Lattice Space

Implicit in the structure of renormalisation is the idea of a parastichy vector moving down through the axis and having rise $h = 0$ at that point. This process gives us a way to navigate through a three-dimensional lattice space (d, h, J) where the jugacy variable only takes integer values. We saw in the previous chapter that the principal vectors of a tree with parameters $(d/J, h/J, J)$ were J times the principal vectors of a tree with $(d, h, 1)$, so that the van Iterson tree of a multijugate is structurally identical to the $J = 1$ version, but simply reduced in scale by a factor J. Figure 5.10 gives one visualisation of this three dimensional parameter space, but other visualisations are possible.

5.8 Notes to this Chapter

Modelling of phyllotactic patterns as exact lattices began with Schimper [100], Braun [15] and the Bravais brothers [18] in the 1830s; it is not a coincidence that in the following decade Auguste Bravais became a founder of crystallographic theory. But the dominant nineteenth century German botanists, Sachs and Hofmeister, found it impossible to reconcile the crystal precision of lattice theory with actual plant forms, and it was not part of mainstream theory. An exception was Schwendener's 1878 publication [107] modelling organ placement as a stacking of decreasing size shapes including disks.

This idea started to appear in textbooks in both English and German at the beginning of the twentieth century [131], and motivated van Iterson's 1907 PhD thesis on

5.8 Notes to this Chapter

transitions through arrangements of fixed size disks on the cylinder, leading to the van Iterson tree of touching-circle lattices [127]. As Christophe Golé has pointed out, van Iterson actually went further than this, calculating how to place disks on cones and planes in analogous lattice-like patterns. A special case of this is the mapping from straight-line parastichies on the cylinder to logarithmic spirals on a disk that we will see in Chap. 6, thus leading naturally to models in which disk sizes vary naturally with position.

At around the same time, Church took a different strategy [24]. By analogy with electromagnetic lines of force, he argued that parastichies intersected at right angles, and thus modelled patterns as (in our terminology) only square lattices, albeit possibly with later stretches. The 1940s saw much new information on the mechanisms of node placement, thanks to the experiments of the Snows. Richards drew on this, and the ease in which realistic sunflower-like patterns can be constructed without insisting on square lattices, to reject both Church's framework and to half-reject the idea of a fixed global divergence angles. Instead of each point being placed in the lattice according to a repetition of the divergence, he proposed, under the influence of Schoute and the Snows, that each new node would be placed between the nearest two preceding. Richards proposed that the placement would be so that the angular distance between the two older ones would be divided in the golden ratio, or that the resulting parastichy lines would be orthogonal. He appeared to think, incorrectly, that these two conditions were the same.

van Iterson's work was known to Turing as a models of phyllotaxis [124, Sect. 14]. Turing believed that combining a touching-circle lattice with a closest-packing rule would satisfy the Hypothesis of Geometrical Phyllotaxis but found this approach 'unlikely to be valid', perhaps because he knew that empirical lattices are usually not touching-circle lattices. In the absence of empirical data on the actual node-placement function in developing meristems, this objection can't be completely discarded, but Turing's own theory, and other simpler and biologically plausible node-placement functions are quite consistent with touching-circle lattices during development if not in the mature plant. When van Iterson himself returned to the subject after an interval of 53 years [128] he concentrated on node-placement functions based on lattice formation through cylindrical harmonics and thus with an affinity to Turing's reaction-diffusion model.

By the 1970s, Mitchison's view in *Science* was that Richards' work [97] had nearly, but not quite provided a key to explaining Fibonacci phyllotaxis, and that a 'somewhat elaborate' mathematical theory had been proposed by Adler [1]. Mitchison's own paper [80] appreciated the significance of the van Iterson tree for understanding Fibonacci phyllotaxis; it implicitly but incorrectly assumes that there is only one branch of solutions down from the triple-point, for it provides no proof that only one branch is most close-packed. What Mitchison did show numerically was that for a specific inhibition based model this is true, saying in our notation that for this model 'the important point is that x_{m+n} does not stray across the axis as h decreases; in fact it is displaced in the opposite direction.' Erickson [38] emphasised orthogonal parastichies over principal ones and found a decomposition of lattice space by square lattices.

At the start of the 1990s, physicists began to take an interest in the problem. Both Douady and his colleagues [34] and Levitov [73] (after discussions 'with A Sidorov in Moscow') separately studied node-placement models as soft-edged attracting disks, and both noted the connections between the structure of their results and the van Iterson tree. Similar results were obtained by Kunz and his co-workers [68]. The final piece of the Standard Picture was published by Douady [32]: that the packing efficiency is lower on the nonopposed than the opposed branch below a triple-point of the van Iterson tree. The other mathematical papers in the same edited volume for the most part show convergence to the Standard Picture [61].

Levitov [73] was the first to point out the existence of the renormalisation transform in the context of phyllotaxis, and its consequences were worked out by Atela, Golé and Hotton [8]. The results of Sect. 5.4 show, in hindsight, why the van Iterson partition of the upper-half plane is isomorphic to the decomposition of the hyperbolic disc by the modular group of Möbius map that M. C. Escher indirectly relied on for his well-known prints. This decomposition was sketched by Gauss in the context of solutions of a hypergeometric equation and then worked out by Riemann, Schwarz and especially Poincaré in the second half of the nineteenth century [45, 88, 105, 112]. In the twentieth century, one development of this idea to higher lattice dimensions led to what Minkowski called the geometry of numbers and then into modern geometry and number theory [11]. My use of the signature of the Euclidean decomposition to explain the otherwise mysteriously unexplained Δs that pepper Jean's book is the first I have seen in phyllotaxis.

The relation between the Euclidean algorithm coefficients and the van Iterson coefficients I sketched in Sect. 5.4 can though be found half-buried across the pure mathematical literature [26]. Chapter 20 of Schwartz [104] illustrates the relationship between Möbius transformations, continued fractions and Farey intervals in a way that nicely illustrates the structure of the van Iterson tree of Chap. 5. Katok [64] shows how the labelling of geodesics by their continued-fraction, or equivalently Euclidean-reduction, signature that underlies the structure of the van Iterson diagram can be specifically traced back to Artin in 1924 [5].

5.9 Summary

The key result of this chapter is Fig. 5.5. In summary we have found that the space of lattice parameters (h, d) contains a tree of touching circle lattices, which we called van Iterson solutions, which can be labelled by their two shortest parastichy vectors as m=n. This tree bifurcates at triple-points $(m, n, m + n)$ where the lattice is hexagonal and the shortest parastichy vectors $\mathbf{p}_m$, $\mathbf{p}_n$ and $\mathbf{p}_{m+n}$ all have the same length. We can navigate through the tree by decreasing h and we saw this was the same as decreasing the radius of the touching circle. If we navigate down the branch m=n, then it splits at the triple-point into one van Iterson branch $(n, m + n)$ on which the radius continues to decrease and on which the principal pair is opposed, and one other van Iterson branch on which $h(d)$ is flatter and, although the radius

5.9 Summary

also decreases, the packing efficiency at a given radius is smaller. On this branch the principal pair is not opposed, at least at first.

It is not enough for an explanation of Fibonacci structure to find a model that enforces touching-circle lattices, for those make up all of the branches of the van Iterson tree, with and without Fibonacci structure. Turing (and van Iterson) saw that that was needed to generically explain Fibonacci structure in a wide class of models was to find a common reason for consistently choosing the $(n, m + n)$ branch. Turing christened the necessary assumption the *Hypothesis of Geometrical Phyllotaxis*. As Turing writes [124], 'the hypothesis is itself quite arbitrary and unexplained' and why it should hold is a question that 'the geometrical approach is not capable of answering'. We have classified some model equilibria: but we need to know whether they are stable or not and so we need in the following chapters to study dynamic models for node placement. But clearly any mechanism that, say, preferentially chooses opposed over unopposed packings, or more plausibly which preferentially chooses more closely packed lattices, will be likely to generate Fibonacci structure.

We used three different approaches to understand lattice space: an algebraic slog, a bifurcation theory approach based on deformations of hexagonal lattice, and a more sophisticated technique based on renormalisation maps from $(1, 0)$ lattices. Each have their advantages, but all are limited in the extent to which they can be generalised to non-exact lattices.

Nevertheless, these analyses start to provide deep explanations of why patterns of soft, packed, disks, arising from models with long-range attraction and short-range repulsion between nodes, might naturally form Fibonacci structure under changing parameters. Our aim for the rest of the book is to ask what we can infer about biological processes from the observation of Fibonacci structure—or the lack of it. To do this we will need to study a range of models which we will shortly turn to in Part III.

Chapter 6
Transformed Lattices

Abstract This chapter collects the mathematical tools needed to transform lattices without preserving isometry. The lines drawn on the sunflower, seen locally, join the nearest seed positions and they are by definition the shortest distances between those two points. Yet globally they are not straight lines but curved spirals. Conversely, if we take the lattices of previous chapters and project them onto two-dimensional surfaces which are neither planar nor cylindrical, then there will be metric distortions: we have to stretch the parastichy lines. As we saw by thinking about increasingly squashed lattices, with fixed divergence d but decreasing rise h, the definition and perception of the principal parastichy numbers depend crucially on the metric properties. The same lattice projected onto two different surfaces will typically have the same sets of generating pairs but the specific pair with the property of being the shortest can change depending on the projection. This is to some extent a mathematical artefact: biologically there is no initial constant-curvature surface of sunflower meristem cells upon which organ commitment actually happens before being rolled into the final shape. In practice, each such 'lattice point' is a developmental commitment made at a particular times and geometry of the surrounding tissue micro-environment.

6.1 Smoothly Changing Lattices

The bifurcation theory approach of the previous chapter invites us to think about the paradoxical idea of the 'smoothly changing lattice'. We saw, for example, that lattices with a golden angle divergence had parastichy numbers which were increasing pairs of adjacent Fibonacci numbers as the rise decreased. Figure 6.1 presents the same result is a different way, applying a non-uniform vertical stretch to a golden lattice. While we can no longer define parastichy numbers using the lattice formalism, they are still easily identifiable visually; Chap. 9 will give a way to formalise this but for now this intuitive visual definition will suffice. If the change of rise from point to point is too sharp, then we will not necessarily see Fibonacci numbers, even in a lattice stretched from one using the golden angle, but in Fig. 6.1, the transformation has been carefully chosen to show a change from (3, 5) through (5, 8) to (8, 13), and

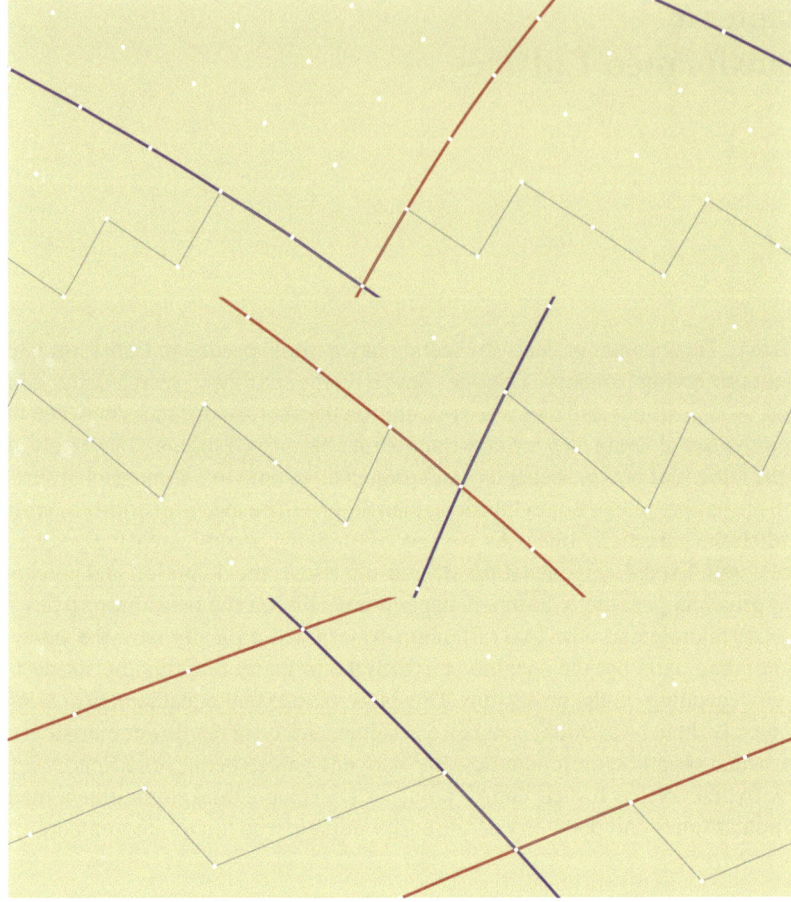

Fig. 6.1 Lattice-like patterns change principal parastichy numbers under smooth geometric transformations. A regular golden lattice has its z-coordinates smoothly transformed so that the rises between successive points become smaller. From bottom to top, the parastichy pairs (3, 5), (5, 8) and (8, 13) are shown. Compare with Fig. 1.3

even here there are some intermediate heights at which assigning a parastichy pair at all is not straightforward.

Although phyllotactic patterns are undoubtedly deformed after creation, it is important to understand that, e.g. large Fibonacci parastichy numbers are not formed in this way empirically. Nevertheless this approach is useful for thinking about the scale by which geometry must change for large Fibonacci numbers to be possible. Figure 6.2 shows how the parastichy numbers for a golden lattice depend on the geometry of the touching-circles making up the lattice: we see that a change of developmental scale of 100–250 fold is necessary to move between a (0, 1) and a (144, 233) lattice.

6.2 Lattices on Disks

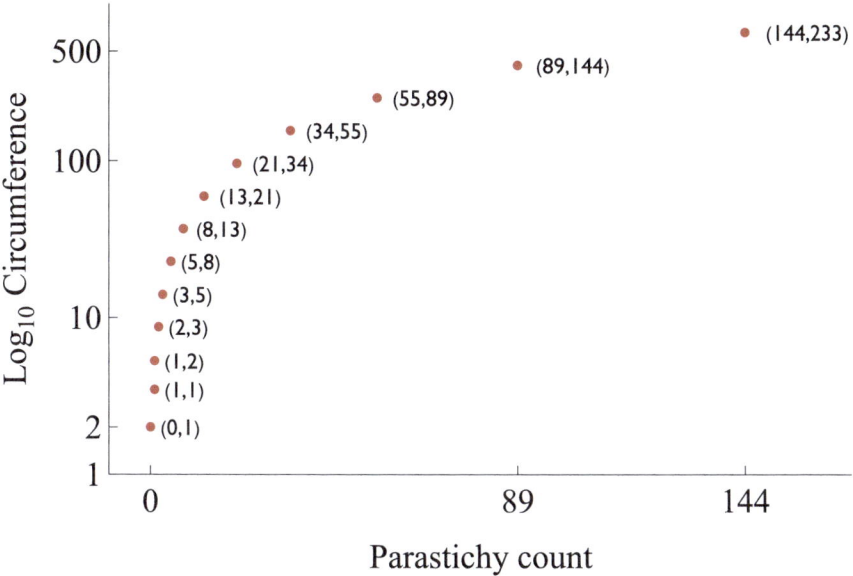

Fig. 6.2 The relationship between the stem circumference relative to the developmental lengthscale and principal parastichy counts. If the touching-circle of a golden lattice is fixed to have radius 1, then Chap. 5 shows that the circumference of a cylinder on which a hexagonal lattice has parastichy counts m and n is $\sqrt{m^2 + mn + n^2}$, as illustrated here for Fibonacci pairs

6.2 Lattices on Disks

A related stretching of lattices comes from putting them onto disks. Visualisations of phyllotaxis on a disk rely on looking down from above on either a microscopic horizontal slice across the SAM, or on a macroscopic view onto the head of a sunflower, but no matter the scale a crucial point is that the oldest organs placed are on the *outside* of the disk. So we would like to construct a map between the lattice on a rectangular cylinder of Fig. 4.2, with rectangular co-ordinates (x, z) and one on a disk with co-ordinates r, θ. If we work on a range $0, z_{max}$ of the cylinder, and then we need a decreasing function of z, $r(z) : [0, z_{max}] \to [1, r(z_{max})]$ to move to in lattice polar co-ordinates $(r(kh), 2\pi d)$ on the unit disk. In particular $r' = dr/dz < 0$.

The difference in radial distance $r(kh) - r((k+1)h)$ is sometimes known as the plastochrone. For h small relative to 1 this is approximately $-hr'(kh)$. We know from previous chapters that even if the divergence angle of a lattice is fixed, the principal parastichy numbers for the lattice depend on the aspect ratio of the strip between the kth and $k + 1$th lattice point on the cylinder. On the cylinder this aspect ratio remains constant and equal to $1 :: h$. But the corresponding annulus on the disk has circumference close to $2\pi r(kh)$ and thickness close to $-hr'(hk)$ and so the aspect ratio is $1 :: -hr'/2\pi r$. The area of the annulus, which is the same as the area per node, is $-2\pi hrr'$. We can find useful functional forms for $r(z)$ by considering how these two quantities scale with r when a cylindrical lattice is mapped onto the disk.

6.2.1 Exponential Scaling

If we control the aspect ratio by setting it as $1 :: h/\alpha(r)$ then

$$\alpha(r) = -\frac{1}{2\pi}\frac{r'(z)}{r(z)}. \tag{6.1}$$

In particular, the aspect ratio of the strip is kept constant by the exponential scaling

$$r(z) = \exp(-2\pi\alpha z). \tag{6.2}$$

The exponential scaling has been a historically common choice for the radial scaling $r(z)$. It maps parastichy lines to logarithmic spirals, and because it preserves aspect ratios also preserves the principal parastichy numbers.

The ratio of the radii between the kth and the kth node is

$$R = \frac{r((k+1)h)}{r(kh)} \tag{6.3}$$

$$= \exp(-2\pi\alpha h) \tag{6.4}$$

and this is sometimes called the plastochrone ratio; assuming that this is fixed is equivalent to assuming the scaling *is* exponential.

6.2.2 Quadratic Scaling

If we supposed the area per seed was $2\pi h \beta(r)$ then we get from above that $-2\pi h r r' = 2\pi h \beta(r)$ and if we further set β as constant get

$$r(z) = \sqrt{1 - 2\beta z}. \tag{6.5}$$

6.2.3 Biological Choices

Figure 6.3 gives an example of the same cylindrical lattice transformed onto a disk with two different functions. The historical attachment to exponential scaling has often, I suspect, been made for mathematical convenience rather than based on biological data, if it has consciously been made at all. Exponential scaling preserves parastichy counts, while quadratic scaling preserves seed size and does display parastichy transitions. Since many sunflower heads do show elements of both, it might be tempting to use a functional form for $r(z)$ intermediate between exponential and quadratic scaling, together with a golden ratio lattice, to yield a model sunflower.

6.2 Lattices on Disks

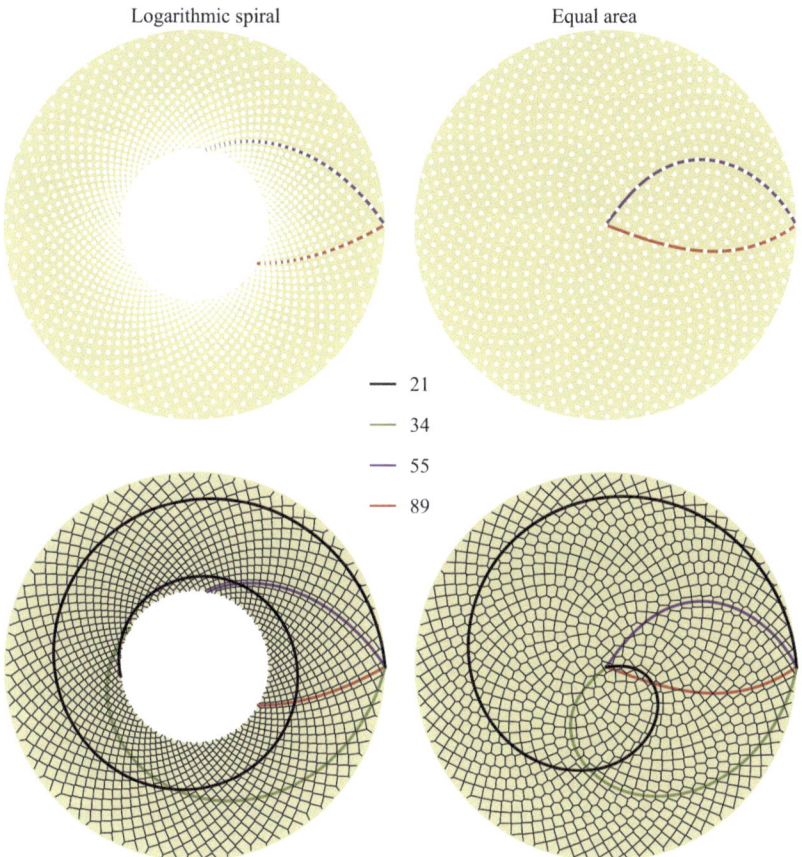

Fig. 6.3 Transformation of a cylindrical golden lattice onto a disk through (left) an exponential transform and (right) a quadratic one. On the left the exponential transform leaves the principal parastichies remain unchanged: the 55 and 89 spirals are the ones that go through adjacent cells both on the outside and the inside of the rim. On the right the quadratic transform gives transitions from (55, 89) down through (21, 34). The exponential scaling is characterised by the parastichy lines becoming log-spirals, but the quadratic scaling by the area of each point remaining approximately constant. On the bottom row the Voronoi cells around each point are plotted. On the left, the z-coordinate of the cylindrical lattice is transformed with $r = \exp(-\alpha z)$, on the right with $r^2 + \gamma z^2 = 1$, with the constants α, β, and γ chosen so that the annulus between the first and second-outermost points has the same area and height:width ratio as the cylindrical lattice. For the exponential transform but not the quadratic one the aspect ratio remains the same throughout the disk

Figure 6.4 is the latest of many examples of doing this. One might even be tempted to try and infer the disk scaling of an observed pattern by fitting such a model. But in understanding the biological process fitting such a form would have little explanatory value. Biologically there is no Euclidean cylindrical lattice to map from in the first place, and even if there was quantifying a map from the geometry and dynamics of

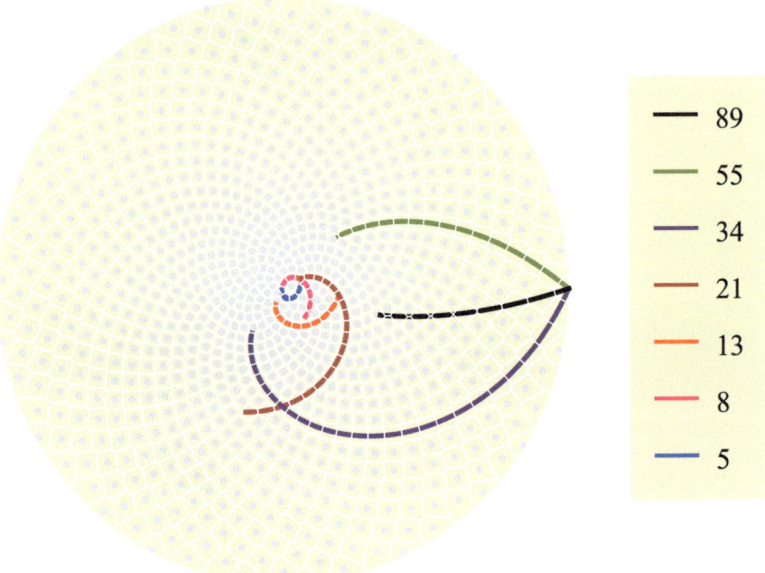

Fig. 6.4 A model sunflower head constructed from a golden lattice together with disk scaling function that allows seed size to become smaller at the centre of the disk, with principal parastichies indicated

the developing plant to that of the observed outcome is not a prospect with much chance of success. Instead we will start to need to look at the dynamics of organ placement on the embryonic seed head.

6.1 Consider a golden lattice, vertically stretched so that the cylinder segments into regions C_k where pattern is close to a lattice with parastichy counts of adjacent Fibonacci pairs (F_k, F_{k+1}). Show that the average rise in segment C_k scales like τ^{-2k}. Find a relationship between k and the plastochrone ratio R of a disk pattern mapped from a near-Fibonacci lattice with an exponential scaling.

6.3 Other Arenas

The map between a cylinder and a disk is a special case of the more general problem of lattice-like structures on surfaces of revolution. One way to map our regular cylindrical lattices to those on a bulging cylinder, that is a surface of revolution described by a profile function $r(z)$ giving the radius of the shape at each height, is to use the vertical arc-length $s(z) = \int^z \sqrt{1 + r'}$. If we interpret this as rise in the original lattice, then each point can be mapped onto this bulging shape. The final image in Fig. 6.5 shows the shape from above and illustrates one way in which

6.3 Other Arenas

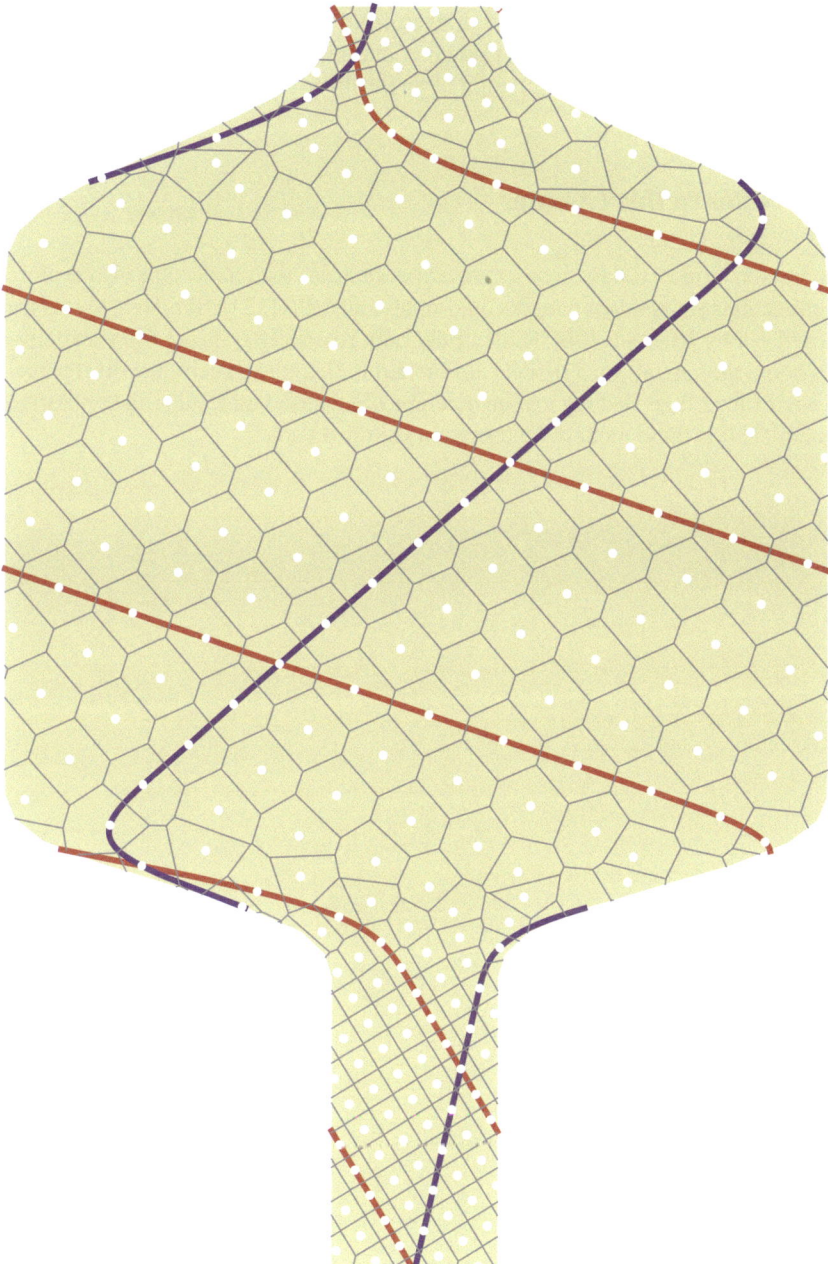

Fig. 6.5 A golden lattice painted onto a bulging cylinder with the (5,8) parastichies highlighted. The lattice is transformed so that the z-height in the planar lattice corresponds to arc-length s in the vertical direction on the bulging cylinder. See text

a cylinder-to-disk mapping can be derived from a shape function. But again it is important to bear in mind that in practice the patterns emerge on the developing shape, not through the transformation of an idealized cylindrical lattice.

6.4 Notes to this Chapter

van Iterson himself did not restrict his lattice analyses to cylinders: he also explored, in effect, mapping lattices patterns onto cones and disks [127]. Similar analyses were carried out by Richards [96], Ridley [98], and Yeatts [133]. From the viewpoint of this book, though, all such models are limited in that they specify the divergence in advance rather than viewing it as an outcome of localised molecular node-placement dynamics. It is these models we turn to in the third Part.

Part III
Mathematical Modelling

Chapter 7
Developmental Biology of the Plant Stem

Abstract This chapter provides a mathematician's guide to the developmental biology of the plant stem. Plants have flexible but tightly controlled developmental mechanisms for generating new organs in the larger spaces left by previous ones and this generates pattern. Many of the molecular and genetic control processes are quite well understood, particularly in the model organism *Arabidopsis*. This will allow us, in Chap. 9, to recognise the basis of a number of different primordium-placement models.

7.1 Stem Extension and Thickening

There is some evolutionary basis for expecting that stem morphogenesis is based on the same fundamental mechanisms across a wide phylogenetic spectrum. The first plants arrived on land around 470 million years ago [53, 57]. They were leafless, without stems, and reproduced through spore distribution. The plant stem evolved relatively quickly around 40 million years later, probably driven by the selective advantage of distributing spores over a wider area. The molecular mechanism that allowed stem production seems to have evolved only once, but was so useful that it soon after was associated with an explosive radiation of species. There is some suggestion that very early plants did possess some form of regularity in their architecture, and by the time of the first angiosperms there was a central and strongly conserved role for auxin in the patterning mechanism [94].

The stem growth mechanism in vascular plants is now well understood at a genetic level. Broadly speaking, control over cellular division and commitment is exerted by the shoot apical meristem (SAM), a relatively small region of cells in a bulge over the top of the stem itself (Fig. 7.1). During stem growth, cells at the lower rim of the SAM and in the outer two or three layers of the stem are prompted to differentiate (Fig. 7.2), lifting the 'roof' of the SAM, and causing the more internal cells, labelled as rib meristem in Fig. 7.1 to proliferate so as to resolve the resulting mechanical tension. It is near the border of this continuously regenerating rib meristem region and the central organising zone that pluripotent cells are maintained and from time to time commit to primordia formation.

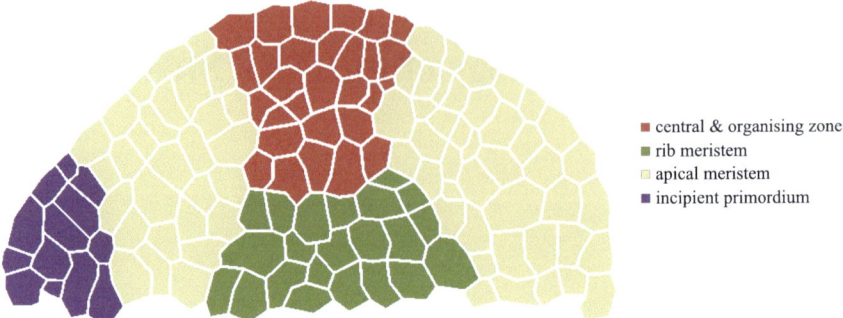

Fig. 7.1 Cross section through an idealised shoot apex, with a central control region, a more peripheral apical meristem which contains regions of cells that are capable of committing to primordium formation, and an incipient primordium in the process of developing into specific organ such as a leaf. Redrawn and modified from [42]

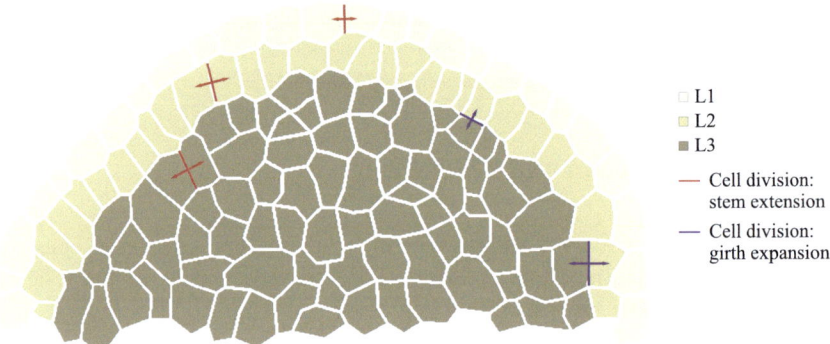

Fig. 7.2 The same shoot apex as Fig. 7.1. The outermost two cell layers, usually known as L1 and L2, preserve a distinct structure and function over development. In particular cells within L1 and L2 derive from anticlinal (i.e perpendicular to the surface) cell division within the layer, associated with extending the stem. By contrast cells within L3 arise from both anticlinal division within L3 and periclinal division from L2 or L3 into L3 which contribute to girth increase. In *Arabidopsis*, primordium initiation occurs from cells in L1

A variety of computational frameworks of varying ambition have now been developed for plant morphogenesis [60]; it is not the goal of this chapter to review these. Instead we want to show how their common geometric features influence pattern development.

Intriguingly, similar patterns can be found outside of the plant kingdom. The seaweed *Sargassum muticum* is a macroalgae which evolved independently of the plants, and auxin plays little or no role in phyllotaxis in this species. Yet the macroscopic branching patterns of bud formation exhibit the familiar 137.5° divergence [74].

7.2 Developmental Commitment of Primordia

As the stem extends, the plant has to decide when and where to commit some of the newly created cells, just at the lower rim of the SAM, to become new organs: we call these newly committed cells *primordia*, and the position of the corresponding organ in the mature plant *nodes*.

It is representing this initial commitment decision as a primordia placement function which is the central task of model-building for mathematical phyllotaxy. We will assume radial symmetry, so that the process takes place in a region at the top of a cylinder which is extending vertically and perhaps thickening radially, and the position of the p-th primordium can be written as $\ell_p = (x, z)$ where x represents a circumferential distance around the cylinder and z the height: the notation of Chap. 4 is consistent with this for the special case when the ℓ_p make up a lattice with a regular rise and divergence.

It is difficult, even today, to make repeated non-destructive observations of the developing SAM. A crucial assumption made in this book is that angular position of nodes in the mature plant reflects angular position at the point of commitment. Though stems can acquire systematic helical twist during growth, there appears to be little or no evidence that angular rotation about the growth axis occurs locally: once placed, nodes do not rotate relative to their neighbours. Similarly, while considerable extension occurs during the vegetative growth of the stem, this is uniform around the stem cylinder, so that a node created earlier in development than its neighbours typically remains relatively lower on the stem. In the case of a composite seedhead seen from above, outer seeds were placed by the plant before inner ones were. These two assumptions together allow data on the temporal pattern of primordia in the plant embryonic region to be inferred from the macroscopic spatial form of the nodes in the mature plant. Figure 7.3 shows how a transverse section across the growing bud at around the level of the SAM can reveal the relative divergence angles of each of the successive primordia. In this figure, the oldest primordia are those with the highest numbers. The shoots from each primordium are growing in the same direction as the main stem so that they encircle and enclose the SAM.

With the advent of higher quality microscopy, enabling inspection of the SAM and of primordium formation, Hofmeister developed around 1860 the rule that primordia are placed at the least crowded spot on the meristem. Much later, in the 1940s, came *Snows' rule*, that primordia form as soon as they can but only when and where there is enough space for them.[1] These rules can be tested by ablation experiments, removing previously created primordia and observing how this modifies the primordium-placement decision. These experiments, over the course of a century, provide support to the principle that it is only the set of most recently formed primordia that determine the positions of the next [95, 111] and that this is because they are the most adjacent.

---

[1] The apostrophe here marks this as a concept attributed jointly to both Mary Snow and George Snow.

Fig. 7.3 Transverse and longitudinal section of a growing bud of the bridewort *Spiraea salicifolia* from van Iterson [127]

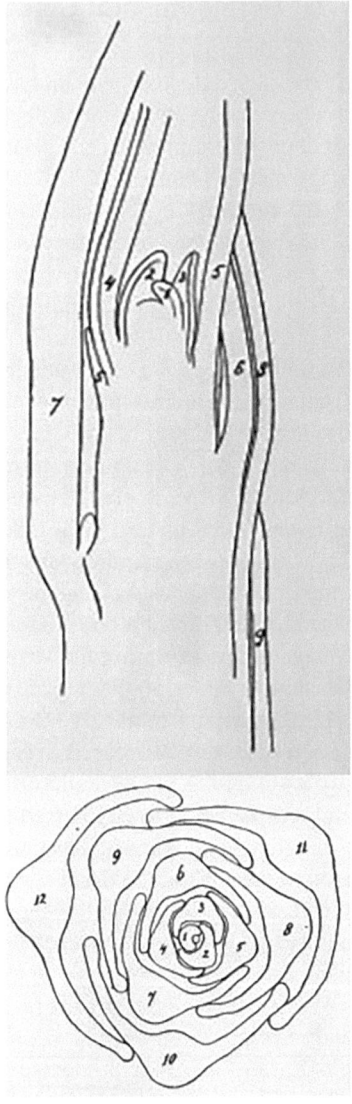

One parameter of geometric interest is the girth of the SAM. It is common for the diameter of the SAM to broaden during the plastochrone, that is when new primordia are not being formed, and then to narrow again at the times of primordium formation, to create an oscillatory diameter as a function of time [29], but I have not been able to find many quantitative studies on how the diameter of the SAM varies over longer course of plant growth. There is some evidence that, in, e.g. sunflowers, the SAM does significantly increase in size as the plant reaches the stage when it is committing cells to form the outer edge of the composite seedhead, and then contracts again corresponding to development of the central regions of the head [86]. Alternatively, the SAM might remain fairly constant in size over the life of the plant, and any stem girth thickening happens later, in older, lower, portions of the stem. It is likely that both processes combine to yield macroscopic changes in stem girth.

7.3 Cellular Architecture of the Shoot-Apical Meristem

Figure 7.4 shows the cellular structure of the SAM in *Arabidopsis*. In simpler plants, there may be only a single cell in the control zone, but in more complex plants like *Arabidopsis* the SAM is multicellular; the dimension of the overall SAM can vary widely from 50 μm to 3 mm in the case of palms and large cacti. The relatively small numbers of cells poses a modelling choice: whether to treat plant tissue as a continuous substrate for biochemical or mechanical dynamics, or whether to explicitly recognise and model cellular architecture. As technology for the latter has advanced, generations of finite-element type models have become increasingly powerful. Identifying accurate cellular architectures is a significant problem for these models, and they have so far been limited to the highly characterised development of *Arabidopsis*.

7.4 Molecular Phyllotaxis

A few gene mutations which cause phenotypic changes in phyllotaxis have been identified, but they have proved hard to interpret to understand primordium placement [67]. Instead, understanding of phyllotaxis has instead largely come from spatial imaging studies of tagged proteins and reporter genes. The processes that control the size of the SAM region and the primordium-placement function are increasingly well characterised at this molecular level, at least in the model organism *Arabidopsis thaliana* [29, 44].

A central role of auxin in self-organised patterns has long been suspected. Auxin plays many other roles in the plant, notably in shoot tropism, but its significance for primordium development was confirmed at a molecular level by Reinhardt and his colleagues [95] and this, together with a modelling component was by 2006 forming a 'plausible model of phyllotaxis' [110]. Recently a consensus has emerged about many though not all of the details of how auxin drives stem node formation [108].

Broadly, pattern formation results from local suppression of the auxin response in the SAM, a suppression controlled by the mutual inhibition between the *KNOX1* and *ARP* gene families. *KNOX1* is widely expressed in the SAM, but primordium commitment requires *ARP* expression. Auxin transport is controlled at the cell wall by proteins from the *PIN1* and *AUX1* gene families. Low levels of auxin within the interior of the SAM promote SAM expansion, and that these auxin levels depend more on auxin transport below the epidermal layer [109] and may lead to more complicated auxin flux patterns, which can be different in the different tissue layers.

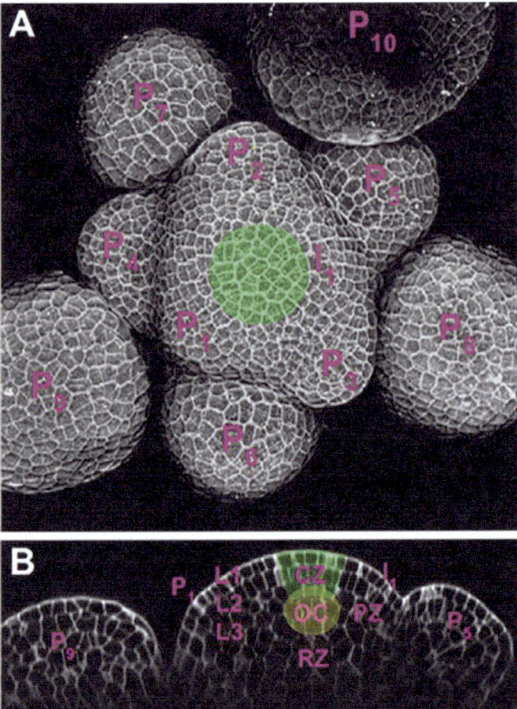

Fig. 7.4 Cellular structure at the SAM of *Arabidopsis*. P_1 to P_{10} label recently generated new primordia, with P_{10} the oldest, illustrating a $(3, 5)$ parastichy where, e.g. P_1 is in contact with P_4 and P_6. Note that the circumference of the SAM, which is roughly indicated in green, is of the order of 40 cells and each cell is very roughly 5 μm at its widest. The transect view below shows two single cell layers L_1 and L_2, which in higher plants will form the epidermis and subepidermis of the mature stem and which divide only 'anticlinally', that is down the axis of the stem in a lengthening manner, while the region L_3 will develop into vascular tissue and can divide in all directions. Recent evidence is that the mechanics of auxin flux differs between these layers. ©Elsevier, from [108]

Low levels of auxin within the interior of the SAM promote SAM expansion, and these auxin levels depend more on auxin transport in L3 than in L1 and L2 [108, 109].

Some recent work [44] has shown that PIN polarity directs auxin flows to both the centre of the SAM and also the most recently created primordia. These authors also suggested that the third to fifth most recent primordia acted as auxin production centres, and suggested a time-dependence in the development of auxin levels after primordium commitment through mechanisms which are as yet unclear.

Early flow models assumed that cells respond to auxin flux by increasing their transport capacity in the flux direction. Although these 'with-the-flux' based models are still believed to work well to explain sub-epidermal protein patterning, they have been challenged as explanations of pattern at the epidermis. If instead the cell concentrates its PIN near the cell membrane of one of its neighbours that already has

7.4 Molecular Phyllotaxis

the highest auxin level, then it is possible to have an 'up-the-gradient' transport. The relative importance of these two mechanisms is not clear: as van Berkel et al. [12] concluded, 'all current models explain *in planta* auxin and PIN patterning to this same limited extent'. A related point of recent controversy and relevance to modelling is what contributes to observed decreases in auxin concentration within new primordia. Conceptual models from the 2010s suggested that auxin was actively degraded, but more recent experimental data has been used to argue that it is transport that removes high auxin concentrations from these areas.

Empirical data at this cell-membrane level poses a particular challenge to current imaging technologies, and while there is active biological and computational research in this area much remains unknown. Another active researcher, Cris Kuhlemeier, argued in 2017 that the coexistence of these 'competing theories ...for the underlying molecular mechanism emphasise the central fact that we don't really know how the patterning is driven' [67].

Another important hormone is cytokinin, long known to have a role in the maintenance of the SAM itself. It has been interpreted as providing noise-reduction to phyllotaxis, through a proposed mechanism in which a mutual inhibition of cytokinin and *AHP6* oscillates between successive primordia in such a way as to stop the simultaneous formation of primordia at different divergence angles at the same time [13]. Both the effect of auxin on developing nodes and the way in which they in turn affect auxin flux through, for example, vein development are increasingly be understood in three dimensions [30].

The existence of reliable molecular probes for auxin concentration and PIN transport will create considerable new opportunities for pattern modellers. Although the tools used need further validation, Fig. 7.5 shows how it may now be possible in principle to quantify auxin transcription around the outer edge of the SAM. It is also possible to detect auxin directly through the DR5 reporter system. DR5 has, for example, been introduced into transgenic strains of *Gerbera hybrida*, a relative of the sunflower, which now allows classic ablation techniques to be combined with auxin imaging as for example Fig. 7.6.

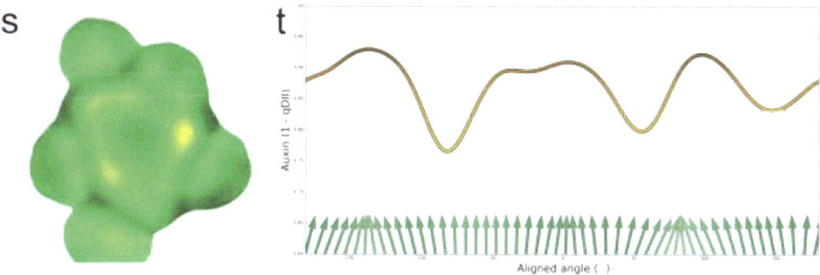

Fig. 7.5 Levels of auxin transcription, measured with DII-VENUS, at radius 35 μm in a developing *Arabidopsis* phyllotaxy. Green arrows show the estimated polarity of PIN1-controlled auxin efflux from cells. Copyright CC-BY-NC-ND the authors, from [44]

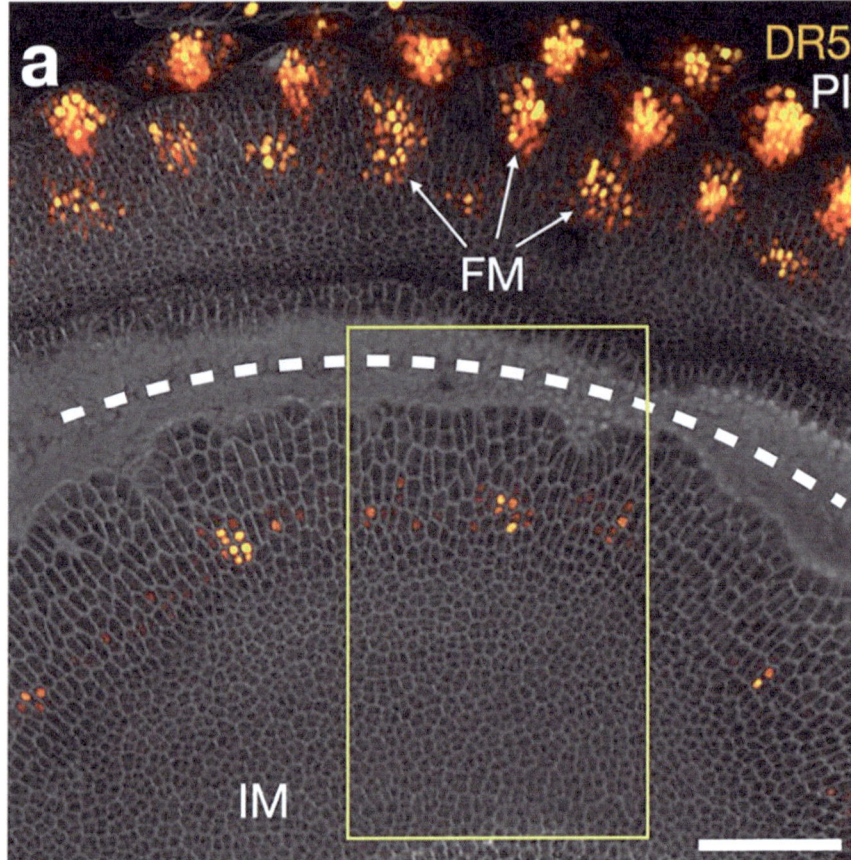

Fig. 7.6 Auxin presence as reported by DR5 imaging in a meristem of *Gerbera hybrida*. The white dashed line shows the location of wounding leading to a incomplete loss of patterning in later development [138]

Quantification of concentration levels is a harder problem but one likely to resolve in future.

One of the difficulties with molecular techniques is their relatively high species specificity, so that it is hard to reproduce the very wide range of species observations of the earlier botanists, but there seems no reason to doubt that we have the fundamentals of an understanding of the origin of Snows' and Hofmeister's rules. By contrast there is there no known molecular basis for any mechanism that enforces a particular divergence angle between successive primordia irrespective of the existing primordium pattern. While plants have sophisticated molecular timing systems used to synchronise with light variation over the course of a day, and transcriptional delays between primordia commitment and formation probably play a significant role [44] there is also no evidence that the SAM possesses the molecular technology for a pacemaker system which could 'drop' new primordia at fixed times.

7.5 Mechanical Stress

Plant cells are attached to their neighbours by relatively rigid cell walls and are subject to both hydrostatic pressure and mechanical strain. Cells have long been known to respond biochemically to these triggers, since primordium commitment causes the meristem to bulge. This could be a way in which the primordium-placement function is coupled to the pre-existing pattern. One early study was by Wardlaw [129], based on a long series of experiments in ferns. The visual analogy between the patterns of phyllotaxis and the buckling modes of a stressed rigid plate was noted in the 1990s by Green [50].

Patterning through stress dynamics is a potent mechanism for form generation, but the coupling between this and auxin dynamics is unclear. It might be that stress provides the local feedback between auxin and PIN localisation [54], or that mechanical stresses do not originate but can subsequently stabilise patterning and subsequent organogenesis [67]. Although models have been developed combining both stress and molecular dynamics [87] that exhibit Fibonacci patterning, they have as yet been inconclusive as to whether modelling can contribute to these questions.

7.6 Vascular Bundles and Leaf Traces

Simple primordium formation models assume a cylindrical symmetry to the stem. However all seed plants develop vascular bundles, each a discrete strand combining an outer phloem tube and an inner xylem tube, and with a number of bundles arranged fairly regularly in a ring to form what is known as a eustele. Typically bundles are interrupted at each primordium; these anatomical issues which directly affect auxin flux are rarely taken account of in current models. While it's unlikely these developing vascularisations affect the original primordia placement choice, they do certainly depend on it. Because each new organ must be connected to one of these vascular bundles, their branching structure provides an alternative record of the pattern-formation process. Modellers have not as yet made much use of these further potential sources of data, and there has been little modelling of vascular morphogenesis itself from this perspective.

7.7 The Relationship to the Standard Picture

The bifurcation-theoretic approach of the first half of this book naturally focuses our attention on the very earliest stages of primordia formation as the embryonic stem emerges from the germinated seed. The first leaf-like structures on the emergent stem are called cotyledons and are distinguished in the botanical literature from the subsequent 'true' leaves; they have related but at least partly independent functions and developmental pathways. In particular, the cotyledon structure develops *before* the shoot meristem itself and through distinct mechanisms which bear little relationship to the developmentally later meristem pattern described above [135]. This cotyledon structure could though potentially provide a 180° polarity cue to the later SAM. At present we know little about SAM during the crucial initiating period after the formation of the cotyledons but before the establishment of patterns with the relative complexity of say Fig. 7.4. This is one of the key empirical areas which would need to be resolved before we could convincingly say we had a tested model of Fibonacci phyllotaxis.

The mathematically natural way in which Lucas numbers emerge from the same patterning dynamics as Fibonacci ones, save for a change of initial condition, poses a direct biological question: when does this happen change embryologically? Is this jump in the van Iterson tree more likely to be accomplished during very early development or later, during a substantial non-linear expansion of SAM geometry? Similar issues arise when thinking about transitions to multijugate patterns. And transitions to these whorled patterns pose further mathematical and empirical challenges. We saw in Sect. 5.7 that it is possible to represent whorled patterns as multijugate lattices in which the symmetry is almost accidentally emergent when the rise is zero and the divergence passes through $1/2J$, but it is very likely that in practice there are other symmetry-enforcing mechanisms at work.

7.8 Development of the Capitulum of the Sunflower

Until recently, almost all of our molecular understanding of phyllotaxis has come from *Arabidopsis* and a handful of other model plants, there is a great deal of morphological data that is worth interpreting in the light of the Standard Picture. A naive interpretation of this Standard Picture is that in the sunflower, say, we should observe simple spiral patterns in the leaf placement on the stem, with typically lower order Fibonacci numbers as parastichy counts, and these patterns increase in complexity by transitioning through ever higher Fibonacci numbers, traceable in parastichies through leaf, bract and floret primordia. How does this fit with the observed form of the mature plant? This varies considerably by cultivar, and by environmental influence. But it is common in fact to see paired leaves at the lowest levels, and then a symmetry breaking higher up the stem [25, 70, 71, 89, 101].

Although in principle consistent with the Standard Picture, there is no substantial dataset on parastichy counts on the stem, or even whether spiral parastichies can routinely be detected, but the divergence is typically between 2/5 and 5/13 [24].

When the stem develops into the seedhead, or capitulum, there is a substantial broadening of its diameter. On the underside of this disk, there are typically no mature organs, and then bracts are closely placed around the rim of the disk. This combination of an absence of recent mature organs and an increase in SAM diameter is entirely consistent with recent findings in *Arabidopsis* [109]. It is a clear—but untested—prediction of the Standard Picture that, although no mature primordia survive in this intermediate range, at least during development there could be auxin patterns of intermediate Fibonacci structure observed, and moreover it is possible

7.8 Development of the Capitulum of the Sunflower

these would be reflected in the surviving vasculature. Moving to the bract region, there are few observations on parastichy counts or countability of these bract placements, but there are a few observations of the total number of bracts occurring which do cluster around Fibonacci numbers [76]. Finally the stem surface folds back on itself so that the floret placement becomes on a near horizontal surface.[2]

When the sunflower is two to three weeks old there is an increase in the cell division rate in the SAM, which causes the apex to swell and within 3 more days to produce a dome. This domed SAM generates a series of primordia which will become side leaves or branches. Within a further ten days or so the dome flattens and broadens, and primordia which will develop into the green bracts become visible around the outside. Figure 7.7 shows these two stages: note how the diameter of the dome at about 20 days is already several millimetres, and that this planar disk expands to a diameter of about 5–6 mm, at the point when floret primordia start to appear.

As the receptacle continues to grow, the vacant generative area stays approximately constant. The seedheads comprise ray florets and disk florets: ray florets are male but sterile florets which often function to attract pollinators while the disc florets are the hermaphroditic fertile flowers which will develop into seeds. Some ray florets, which will develop into the large yellow organs which the layperson might call 'petals', appear inside the bracts but on the outside of the ring; disc florets paired with ray florets form the body of the ring and then ray florets form again on the inside. Within the ring, the paired ray and disk florets form at primordia, where the disk floret appears first and the ray floret is greatly reduced in size [56]. Seed placement thus follows the placement of the disk floret primordia, and in the capitulum typically displays falling phyllotaxis in which the parastichy counts decline as we move towards the most central, and most-recent primordium locations.

When the disc florets actually flower, the flowers tend to obscure the patterning of their primordium placement and it can appear that it is only the as-yet-unflowered disc florets that are regularly arranged, but as the floret heads decay the visibility of the regularity of the entire seedhead returns. In some species (such as that on the cover of this book), the inner ray florets are not visible, but more often they can be seen at the centre of the mature seedhead.

It is hard to quantify the exact scale of the developmental arena on which primordium commitment is occurring because of the delay between primordium commitment at the gene transcription level and visible microscopic change. But even in the absence of reliable genetic markers, commitment seems likely to be occurring in an arena of the order of half a centimetre wide. The argument that there is no known global commitment mechanism to coordinate divergence angles becomes even stronger on an arena of this scale. The fact of pattern generation on this scale has also been used as an argument that reaction-diffusion of the kind envisaged by Turing cannot be acting, as the diffusion times across the capitulum are much too

---

[2] A surprisingly common misconception amongst mathematicians is that the pattern has been laid down from the centre of the capitulum to the outside. Of course the outer ring of the disk is developmentally the *oldest*, and floret placement proceeds from the outside of the disk to the inside.

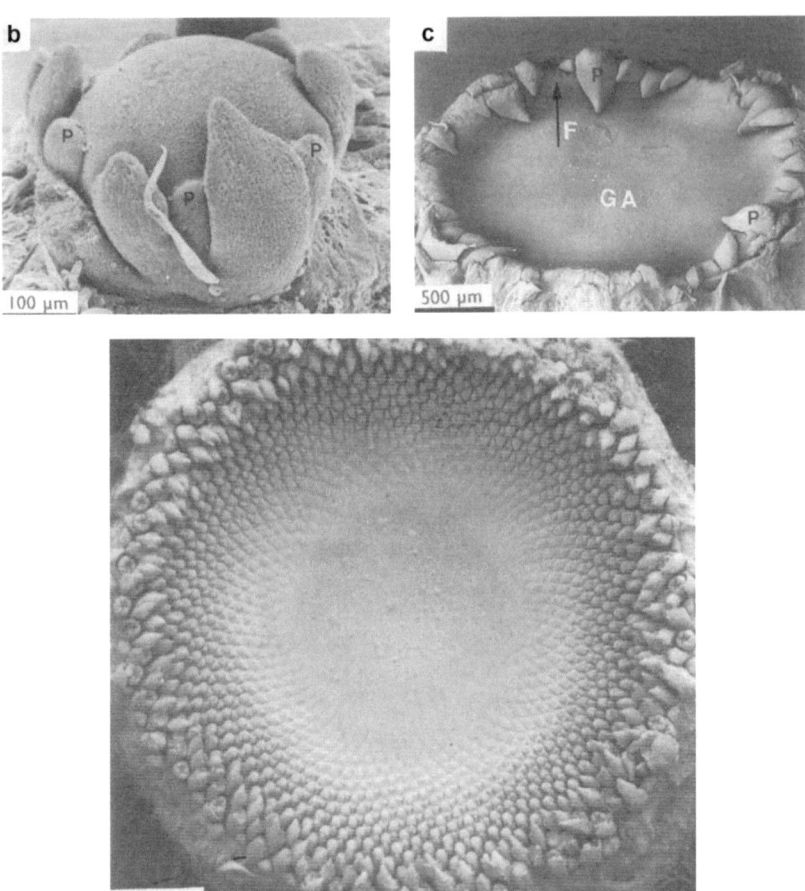

Fig. 7.7 Sunflower capitulum at about 20 days (**b**), when the SAM is generating lateral leaves, about ten days later when the SAM has flattened to a disk (**c**), and then (unlabelled panel below) later floret formation on the disk. Scanning electron reproduced from Figs. 3 and 5 of Palmer [86]. ©World Scientific

long to allow dynamical pattern generation. Although it is unlikely for other reasons that simple diffusion is occurring, this argument is not strong. Even though the particular mechanism is not similar to that hypothesised by Turing in 1952, the core idea that a local primordium-placement function can create global pattern does not require instantly synchronised global information: all that is necessary is that there is a local pattern pre-established over developmental timescales, which we might think of as a morphogen field visualised by the bract placements in Fig. 7.8.

Hernandez and Palmer created a revealing insight by artificially causing an annular wound to the generative area, reproduced in Fig. 7.9 [55]. On the inside of this annulus, primordia production continues but the Fibonacci structure is lost (Fig. 7.10).

7.8 Development of the Capitulum of the Sunflower

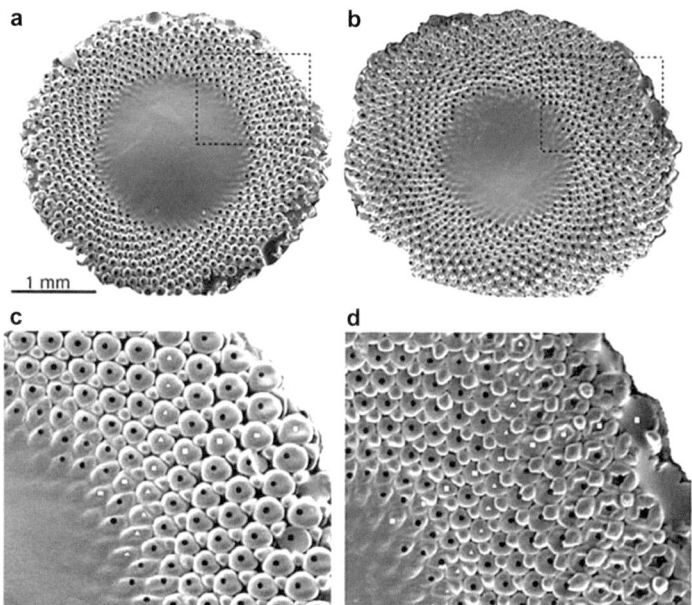

Fig. 7.8 Two successive images (**a**, **b**) of the same sunflower capitulum collected two days apart by casting with epoxy resin. **c, d** Respective close-ups showing how the large ray floret (marked with a black or white dot) and small disk floret are paired close together, and how the primordium patterns, and parastichies, take shape based on the pre-existing pattern at the edge of the capitulum. ©Springer, from [56]

Moreover the first formed primordia on the inside of the wound develop into bracts rather than florets [55].

More recently, ablation experiments have been combined with molecular probes to study patterning in the seedhead in *Gerbera*, another member of the sunflower family [37, 93, 139]. So there is now both morphological and molecular support for the idea that the tissue environment controls primordium formation through an inhibitory effect of earlier primordia which is lost by wounding. The relative uniformity of the post-wounding primordium distribution also supports a local inhibitory effect of primordium commitment. The creation of bracts at the rim of the wound suggests that a commitment to bract rather than floret formation might be a consequence of a lack of local inhibition as well.

100   7 Developmental Biology of the Plant Stem

Fig. 7.9 SEM microscopy of a sunflower head seven and nine days after wounding. Nine days after wounding a pattern of floret primordia has redeveloped. The horizontal bar represents 300 μm. From [55]

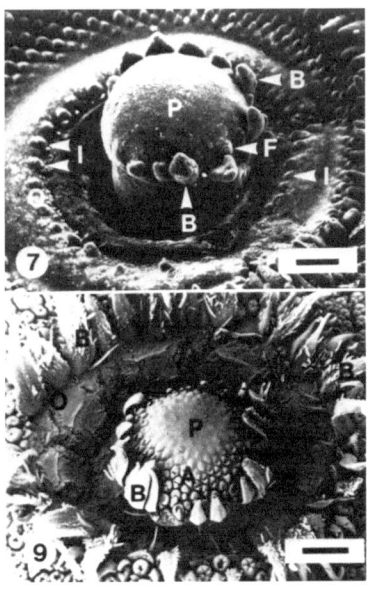

Fig. 7.10 Floret positions of an unwounded (top) and wounded (bottom) sunflower. From [55]

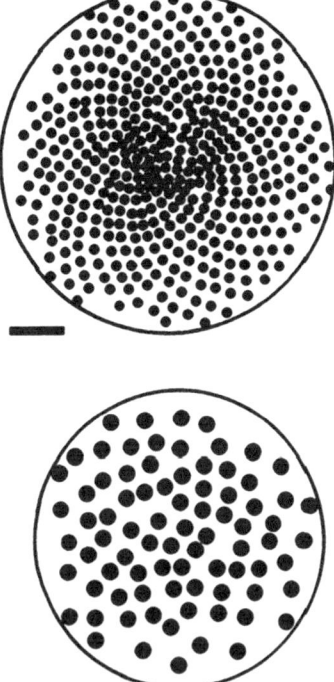

Chapter 8
Statistical Phyllotaxis

> *'most [roses] have 5 petals, but some have 12 or 20, and some a great many more than these'*.
>
> Theophrastus [117] c300 BCE.
>
> *'This observation appears to be off by one. Fibonacci numbers of petals, 13 and 21, are more likely to occur than 12 and 20'*.
>
> Prusinkiewicz and Runions [91] 2012 CE.

Abstract Data collection on the visible patterns of Fibonacci-like phyllotaxis has continued to the present day, although there have as yet been few that combine them with molecular or genetic analyses in the same study. As the outlines of the Standard Picture have emerged, the importance of tracking changing phyllotactic patterns as the plant develops has become clearer. Similarly, the issues discussed in this book about what we don't know, and in particular the hypothesis-testing opportunities of departures from Fibonacci structure, have made the need to identify and survey non-Fibonacci structure more important. Across the wide range of previous studies, there remain considerable data sets which may still be useful for these purposes.

8.1 Early Idealism

I doubt the modern claim in the quotes at the head of this chapter that the ancients were off by one. Prusinkiewicz and Runions don't provide any data to contradict the earlier authority, and I don't know of any reported studies of rose petal counts: there are none from the *Rosa* family cited in Jean's wide survey [60]. But in any case this is a characteristic evidentiary dispute between idealist mathematicians, who know what the answer should be without looking, and empirical scientists who know the answer but think it irrelevant. Cronk has described the early history of plant morphology as dominated by idealists of the former type, natural philosophers who believed that

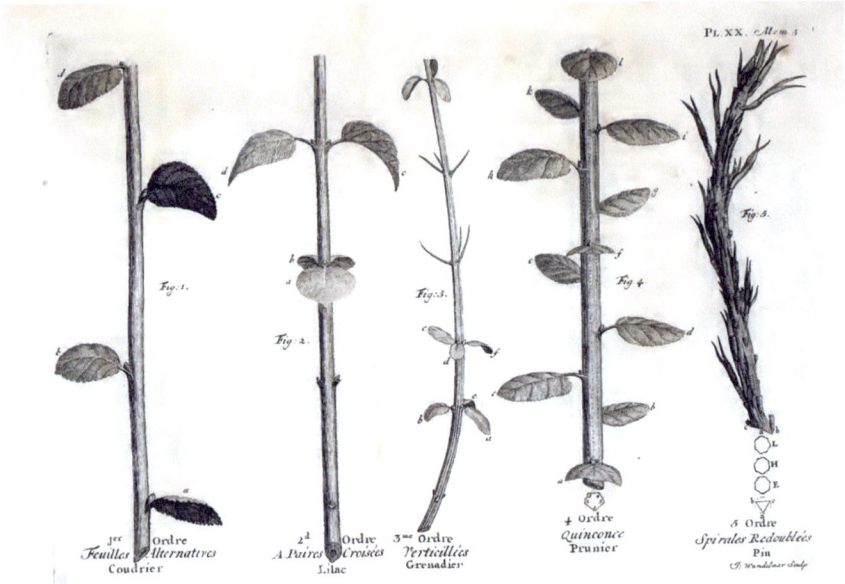

Fig. 8.1 Bonnet's 1754 taxonomy of plant form. Note that in each of the non-whorled patterns the divergence is considered to be rational [14]

reason could identify a small number of ideal patterns that all plants followed. Kepler, for example, had a fascination with the number 5, not least in the structure of the solar system, and saw great significance in its appearance in plant structure [65]. Goethe's belief in the spiral tendency of leaf placement was inextricably linked with a Romantic idea of the unity of life. This idealism was challenged during the nineteenth century by an empiricism, and a turning away from grand theory, of whom the greatest exponents were Hofmeister and von Goebel [29]. Classification of leaf patterns is at least as old as Leonardo da Vinci [36], and became a staple of early modern botanical taxonomy, as for example Fig. 8.1.

The connection between these patterns and the Fibonacci sequence was first made in the 1830s by Schimper [100] and Braun [16, 17], who stayed with the idea of classifying spirals by rational angle and found these were commonly of the form 1/3, 2/5 or 3/8. It seems to have been Braun who first wrote of the appearance of large Fibonacci structure in the sunflower (Fig. 8.2), and also noted Fibonacci structure in pine cones for the first time.

8.2 Observational Phyllotaxis

Data collection on the visible patterns of Fibonacci-like phyllotaxis has continued to the present day, although there have as yet been none that combine them with

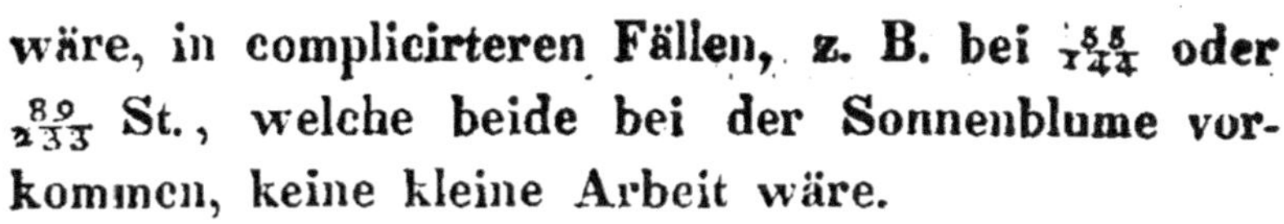

8.2 Observational Phyllotaxis

> wäre, in complicirteren Fällen, z. B. bei $\frac{55}{144}$ oder $\frac{89}{233}$ St., welche beide bei der Sonnenblume vorkommen, keine kleine Arbeit wäre.

Fig. 8.2 Braun's 1835 report of large Fibonacci structure: 'With considerable work, more complex cases such as 55/144 or 89/233 can be found to occur in the sunflower' [16]

molecular or genetic analyses in the same study. As the outlines of the Standard Picture have emerged, the importance of tracking changing phyllotactic patterns as the plant develops has become clearer. Similarly, the issues discussed in this book about what we don't know, and in particular the hypothesis-testing opportunities of departures from Fibonacci structure, have made the need to identify and survey non-Fibonacci structure more important. Across the wide range of previous studies, there remain considerable data sets which may still be useful for these purposes. One of the great strengths of Jean's *Phyllotaxis* book remains its collation of parastichy count data from a wide range of plant species. Indeed *Phyllotaxis* is effectively complete as a signpost to the literature prior to the 1990s, with a handful of exceptions noted by Okabe [85].

In the rest of this chapter I only draw attention to older work that may still be useful for analysis, and some recent papers. Systematically counting phyllotactic patterns from photographs is not that difficult, but it is tedious, and it somewhat surprising that beyond some discrete Fourier transform approaches [82], only a few useful image analysis tools have been published for this task [4]: they would be particularly useful in the analysis of noise and front dynamics.

8.2.1 Statistical Phyllotaxy and Rare Parastichy Pairs

While sunflowers provide easily the largest Fibonacci numbers in phyllotaxis, and thus, one might expect, some of the stronger constraints on any theory, there is a surprising lack of systematic data. Before 2012 there were only two large empirical studies of spirals in the capitulum, or head, of the sunflower: Weisse in 1897 [130] and Schoute in 1938 [102], which between them counted 459 heads: Schoute found numbers from the main Fibonacci sequence 82% of the time and Weisse 95%. Most significantly, neither reported any samples at all which did not have Fibonacci structure. Much more recently a smaller sample of 21 seedheads was carried out by Couder [28] who specifically searched for non-Fibonacci examples, while Ryan et al. [99] studied the arrangement of seeds more closely in a small sample of *H. annuus* and a sample of 33 of the related perennial *H. tuberosus*.

In 2012 I collaborated with Manchester's Museum of Science and Industry (MOSI) [116] to evaluate 768 parastichy counts in sunflowers; of these about three-quarters were strictly Fibonacci numbers, and a bit under a tenth were either double-

Fig. 8.3 Distribution of parastichy numbers in the sunflower head. Left: histogram classified by type of Fibonacci structure: right: magnification of the same plot. From [116]

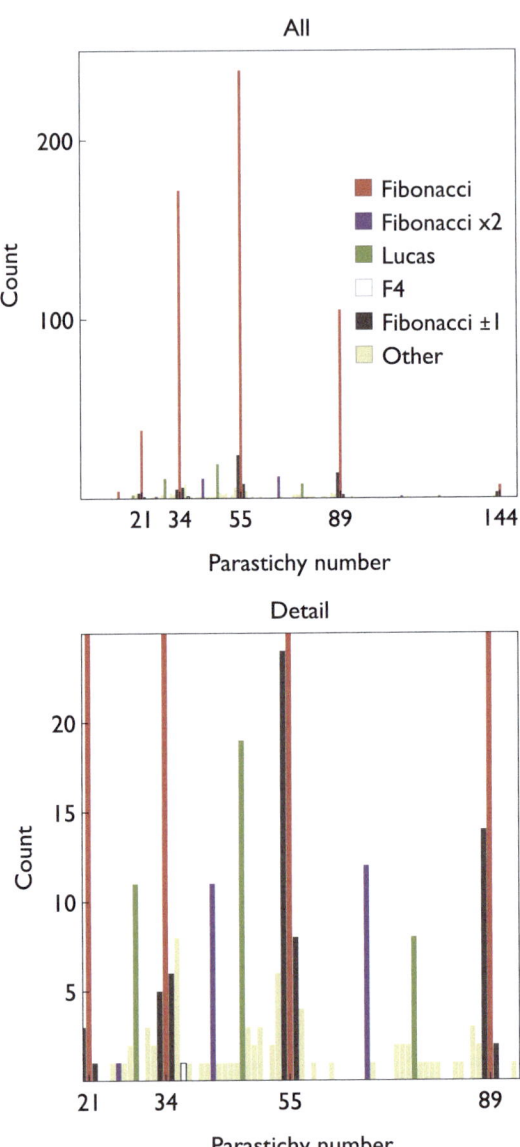

Fibonacci numbers, Lucas numbers, and F4 numbers (Fig. 8.3), so that 'Fibonacci structure' was seen in 85% of cases. But the remainder was a second group of non-strict Fibonacci seedheads. Unlike previous studies, the project found counts that were only approximately Fibonacci (Fig. 8.4). In particular, and to my surprise we found a small but distinct group of seedheads with parastichy counts of the form $(F_n, F_m \pm 1)$, and parastichy counts which were Fibonacci numbers less one were

8.2 Observational Phyllotaxis

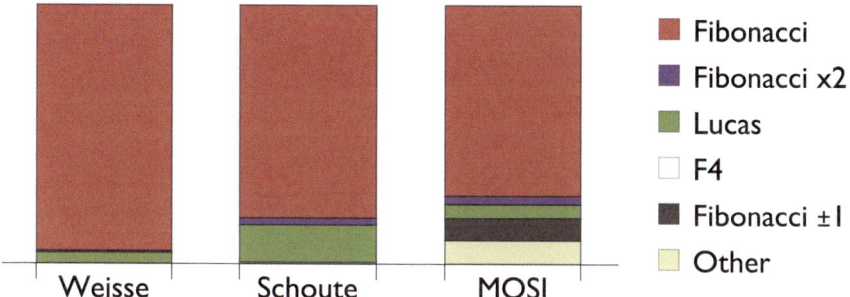

Fig. 8.4 Classification of observed parastichy types. 'MOSI' from [116]; 'Schoute' from [102]; 'Weisse' from [130]. Note that neither Weisse nor Schoute reported any non-Fibonacci structure at all. From [116]

present more often than Fibonacci numbers plus one in a statistically significant manner. Counts of the form $F_n - 1$ were seen more often than ones which were Lucas numbers. The project also reported heads in which no parastichy number could be definitively assigned. Sometimes this was simply due to the poor quality of the specimen, but a number of patterns emerged which the seedhead showed regularity which did not fit into the simple cylindrical lattice paradigm.

8.2.2 Statistical Phyllotaxis of Fir Cones

Fierz single-handedly carried out an analogous study of fir cones, a project with almost ten times as many samples as in the MOSI sunflower study [39]. She found strict Fibonacci structure more often than in the MOSI study (97% of a remarkable 6000 *Pinus nigra* cones), and double Fibonacci were again the second-most common class. Unlike the MOSI study, she did not draw attention to the occurrence of Fibonacci plus or minus 1 counts. In the very small number of cases of non-Fibonacci counts she did see, such as five observations of (8,12), Fierz chose to interpret these as coming from the very weakly Fibonacci-like sequence 4,8,12. While these are small numbers, both of sample size and of parastichy counts, in the light of the MOSI data it is tempting to reinterpret most of her non-Fibonacci observations as small perturbations to Fibonacci counts rather than as the outcome of Fibonacci transitions from exotic starting points. Fierz also helpfully recorded some of unusual, and literally uncountable fir cones which made up less than 5% of the sample. And as the MOSI study threw up the sunflower 667 as a challenge to our pattern classifications, Fierz's cone S5 of Fig. 8.5 also poses a challenge to pattern models. Fierz believed that each of the parastichy counts 7, 8 and 9 was equally good descriptors of the clockwise parastichy counts of this sunflower, and based on the triple-point unfolding we might hope to understand this as a similar transitional pattern. But the

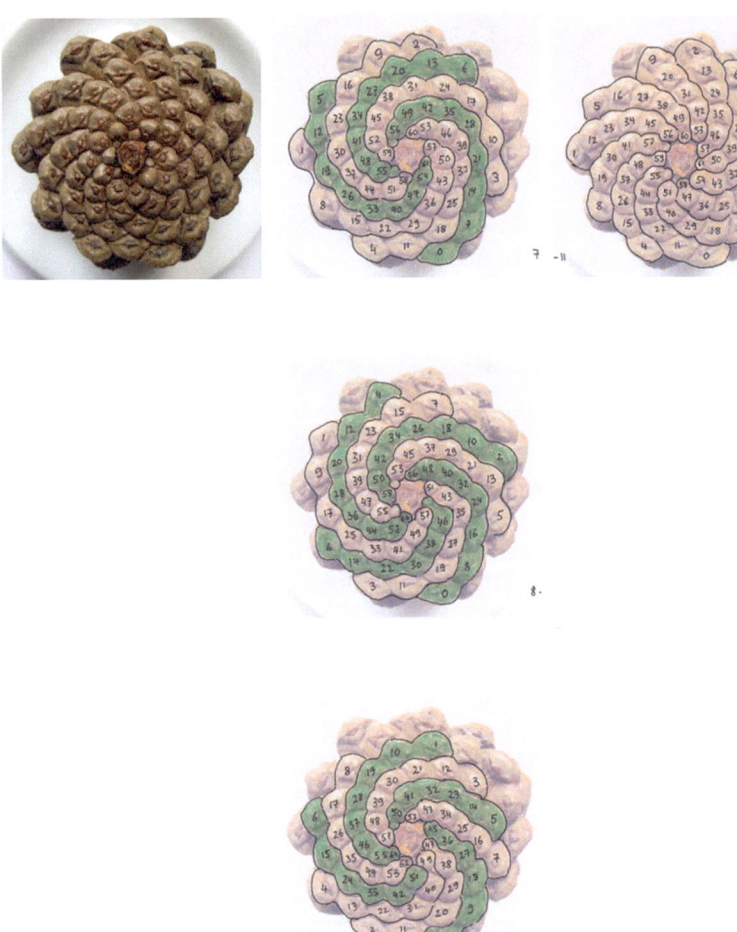

Fig. 8.5 Fierz's cone S5, with equally good clockwise parastichy counts of 7, 8 and 9 and a counter-clockwise count of 11. ©CC-BY Fierz, 2015 [39]

details are still to be worked out, as is whether patterns like these will only be seen if the counter-clockwise parastichy number, here 11, is not Fibonacci.

8.3 Non-lattice Patterns

8.3.1 Loss of Symmetry

One class of patterns not previously recorded en masse was seen on seedheads which had lost rotational symmetry. In examples like sample 667 (Fig. 8.6) the parastichies are locally clear and unambiguous. However set of red parastichies, if continued clockwise, are naturally extended to the green parastichies, while if extended anti-clockwise extend to the yellow ones which have a steeper angle. The result is that it is not possible to provide a clockwise parastichy count for this sample according to the definitions of Chap. 4, although the counter-clockwise pattern is unambiguous. This does not seem to be the same form of non-Fibonacci pattern as the dislocated grids discussed below, but it is highly ordered. For a head close to rotational symmetry, there is a relatively narrow annulus on which transitions between principal parastichy counts occur. As the head becomes more asymmetric, the annulus becomes distorted and in this example it has left the visible seedhead completely. Presumably this can be understood as a lack of rotational symmetry in the node-placement function arising during growth.

8.3.2 Dislocations

Figure 8.7 gives a number of examples from Fierz's fir cone study where she considered that parastichy numbers became un-evaluatable because of a dislocation in the parastichy lines. Zagorska-Marek has classified these types of pattern changes as λ and γ dislocations, taking their name from the letter shapes. In a λ dislocation the parastichy count is reduced, while in a γ dislocation it is increased as in Fig. 8.8. With a deterministic coin-dropping model, γ dislocations are to be expected as coin-size decreases, but they also showed that they can occur as coin-size increases. Even if a node-placement map had reached a lattice equilibrium and was then subject to a small geometric change, it can take a number of iterations before parastichy counts can respond, even if they can be defined during the transition.

The analysis of dislocations in phyllotactic patterns was given new vigour by the theoretical development of cylinder tilings as a unit of analysis, as we will discuss in Sect. 10.4.1. For example Douady and Golé [35] explored empirical examples in fir cones *Pinus pinaster*, *Pinus negra* and *Cedru libani* as well as the bean *Parkia speciosa*.

Fig. 8.6 A set of parastichy lines which don't foliate the sunflower disk. Top image from seedhead 667 of [116]; the original seedhead is now in the collection of the Museum of Science and Industry, Manchester as entry `E2017.2100.1`.

Fig. 8.7 Although over 97% of 6000 *Pinus nigra* cones showed exact Fibonacci patterns, a small number were uncountable by standard methods, because of the loss of a parastichy. ©CC-BY Fierz 2015 [39]

8.3 Non-lattice Patterns

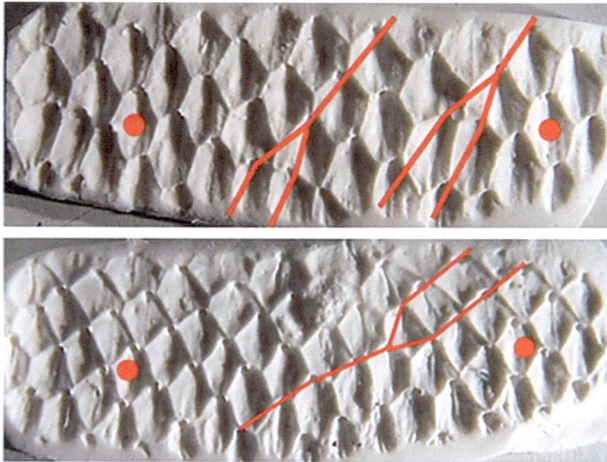

Fig. 8.8 λ and γ dislocations in phyllotactic patterns of *Magnolia acuminata*. In upper clay replica the pattern transitions from (3,8) to (3,7) to (3,6) through two successive *lambda* dislocations. In the lower example the pattern transitions upwards from (4,8) to (5,8) in a γ dislocation. ©Zagorska-Marek and others under a CC-BY-4.0 license, from [137]

8.3.3 Rising and Falling Phyllotaxis

Church pointed out the relative ease of observation of rising phyllotaxis on *Euphorbia* (Fig. 1.3). A strong predictions of the Standard Picture is that parastichy counts should start as low numbers in the older parts of the plant and then increase corresponding to relative SAM expansion. They imply that this holds even as the nodes which are being placed fundamentally change their ultimate functions, from, e.g. leaf to bract to ray floret to disk floret. There have been few systematic attempts to evaluate this in any species.

When we examine the large Fibonacci patterns in *Compositae* there is a difficult missing link in the Standard Picture. There is an observed leap, in the *Compositae* at least, between parastichy counts in the stem and in the disk floret. The common daisy, at least of Bletchley Park, has a handful of leaves, all branching at ground level, whose parastichy structure has never been systematically reported, then a stem which is completely smooth to the naked eye, before the bract, ray floret and disk floret structure appears. That the daisy usually has 13 bracts suggests a transition through 3 and 5 and 8 in the principal parastichies of the node pattern during development: at a minimum that requires 8 nodes, and probably rather more. But are they the ground level leaves, or were nodes created and then suppressed on the elongating stem above them? The situation is even more dramatic in the even more exaggerated seed disk of the sunflower. Sunflower stems do have branches which have been the object of phyllotactic study, but typically reporting (1,2) parastichy counts if any at all. As we have seen, bract counts in the sunflower are not tightly clustered on Fibonacci numbers, though the Standard Picture can explain this away by saying that the bract generation region is not necessarily a tight annulus representing a single front. But even so the Standard Picture does demand that node placement in the bract region relies on a pattern predisposition with at least 13 nodes if not more. Faced with the smooth underside of the disk capitulum, there is no visual evidence of the past trace of such a pattern. Okabe has paid useful attention to the pre-patterns that are seen [84], but there is little systematic data. Of course it is common in development for early commitments to be erased later, but whether this has left more subtle anatomical traces than are visible, or whether the right molecular probe at the right stage of development can yield the missing intermediate parastichy counts, remains to be seen.

8.4 Beyond Parastichy Counts

8.4.1 Bract and Ray Floret Counts

For sunflowers and other *Compositae*, we have seen that if bracts, or ray florets, are generated in a narrow enough growth region, then the Standard Pictures suggest that the count of these organs will be close to having Fibonacci structure, though perhaps less bound to these counts than principal parastichy numbers would be. Moreover these counts should be larger than small parastichy counts on the stem below. Depending on when the geometric shift to a smaller SAM radius occurs, the bract count, the ray-floret count, and then the seed parastichy number should have some detectable relationship. While there have been studies recording bract count and ray-floret count, there has been little attempt to correlate these with lower stem patterning or higher seed disk spirals. In the MOSI dataset there was no clear clustering of either bract or ray-floret counts around Fibonacci numbers (Fig. 8.9). The parastichy numbers were somewhat laboriously but consistently counted in a documented, reproducible manner by the project team, but the bract and ray-floret

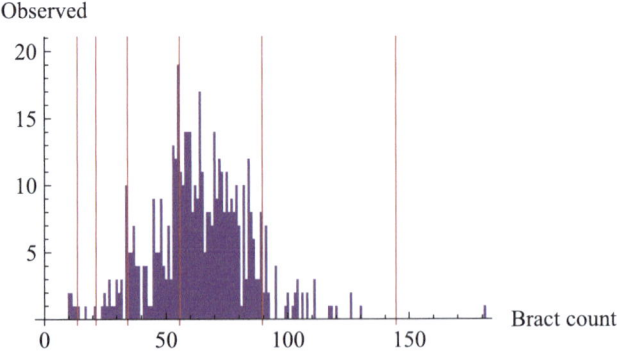

Fig. 8.9 Distribution of bract counts in the MOSI dataset. Redrawn from [116]

counts were as provided by members of the public, and this may explain some of the spread; but equally it might well be a real phenomenon.

8.4.2 Column Patterns

Mathematical phyllotaxis has recently begun to pay attention to other classes of lattice-like structures, albeit ones which have been visible to botanists for rather longer. The sweetcorn cob shows one such pattern in Fig. 8.10; another is visible in many cacti (Fig. 8.11). These show patterns stacked columns of nodes along the vertical growth axis, and there is often a slow twist of these columns with growth. These patterns have only recently attracted significant attention from modellers of phyllotaxis (see Sect. 10.5). These patterns can often display phyllotactic transitions where they insert one new column, such as in Fig. 8.11. That transitions like (5,5) to (5,6) to (6,6) can occur suggests a reason to be cautious of the formalism of Chap. 4.

Fig. 8.10 A commercially sold sweetcorn cob, with column patterns showing parastichy lines nearly parallel to the growth axis

Fig. 8.11 Phyllotactic transitions in an Argentinian saguaro, as reported in [87]. The authors detect a phyllotactic transition from (5,5) to (5,6) to (6,6). ©Elsevier (2015)

Representing a (5,5) pair as a five-jugate lattice $5 \times (1,1)$ is not mathematically inaccurate, but it is wrong to assume from this that all nearby lattices also have five-fold symmetry.

8.5 Summary

There is a now several centuries' worth of individual pattern descriptions as images or as parastichy counts. As the Standard Picture has emerged, the questions asked of this data have changed, with more focus on the information provided by non-Fibonacci or other less structured patterns. Even quite old datasets can still be valuable, but as imaging and data handling technology improves we can expect more focus on rare, but informative patterns.

In any case it is clear from existing observations that sometimes the strict lattice is too rigid a model to describe important pattern properties. In the next chapter we will explore attempts to relax these models to ones that allow more general patterns, and go beyond lattices as a unit of analysis to consider cylindrical tilings.

Chapter 9
Placement Models

Abstract Up to now we have studied only static lattices, which fix the location of every node simultaneously, and then we smoothly transformed the lattice as a whole. This approach cannot represent the one-by-one developmental choices for node commitment, so now we will consider more general node-placement models. This chapter and the next provide a taste of a range of models that have usefully been deployed to understand Fibonacci phyllotaxis in this way.

9.1 The Douady-Couder Ferromagnetic Model

A seminal physical model for phyllotaxis was Douady and Couder's 1992 ferromagnetic system, in which charged drops of oil are dropped onto a cylindrical plate in a magnetic field that causes a centrifugal motion. A caricature of this model is shown in Fig. 9.1. If there is a very long time between successive drops, the drops will move off at random directions; if the process is speeded up slightly then mutual repulsion of the drops will ensure they move in successively opposite directions as in the spots 0, 1 and 2 in part (a) of Fig. 9.2. However if the process is sped up still faster, then the same mutual repulsion squeezes out the new spot to the side. It is this symmetry breaking that has placed the spot numbered 2 away from the line between the two earlier spots in part (b) of Fig. 9.2 and initiated a spiral form.

A model of this spiral generation-phenomenon can be built by assuming that the k-th drop, arriving periodically at the centre of the disk at time $t_k = kT$, has its angle determined as soon as it has left a small inner ring of radius z_c, and then moves off with displacement $z(t)$ so has polar coordinates $⑄ = (z_c + z(t - t_k), \theta_k)$ with $z(0) = 0$, or the cylindrical coordinates above $(x = 2\pi\theta_k, z(t))$. Douady and Couder modelled the radial displacement typically as exponential growth $z(t) = z_c \exp(Gt/T)$. The choice of angle θ_n for the n-th and newest drop is made by minimising $\sum_{i<n} E(|(①, (z_c, \theta_n))|)$, where $|\cdot, \cdot|$ is the distance between two points on the surface of the cylinder and E is an energy function of the form say $E(r) = 1/r^3$. For these and similar functional forms, the system is controlled by the dimensionless parameter G.

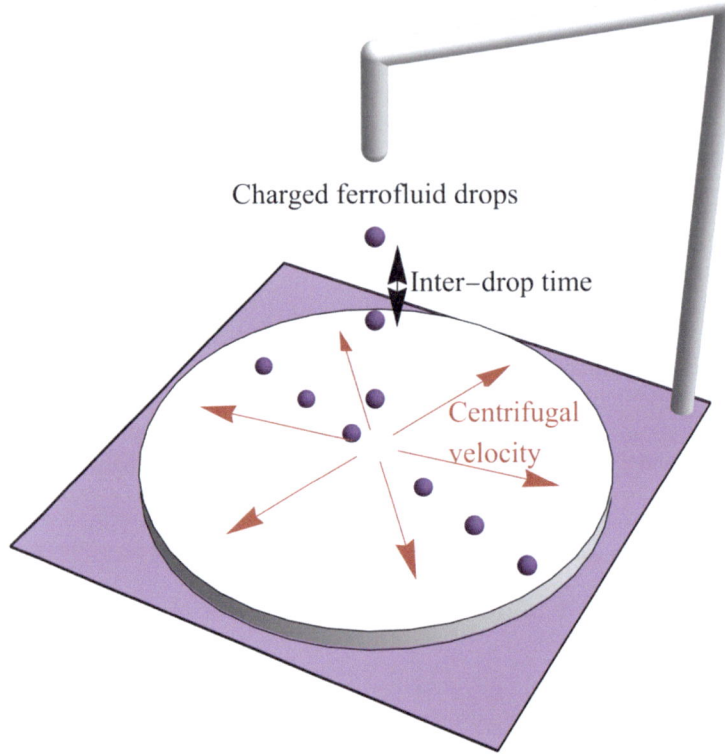

Fig. 9.1 Simplified view of Douady and Couder's apparatus for generating phyllotactic patterns. A centrifugal magnetic gradient transports ferrofluid magnetic dipoles periodically dropped from the central pipette. Redrawn from [33]

Figure 9.3 shows some results of simulation of this model, with a clear path through several Fibonacci transitions. When versions of these results were first published in the 1990s, they formed the first demonstration of a mechanism for Fibonacci phyllotaxis realised in a model which could be plausibly connected to known biology. It was a brilliant proof-of-concept for the possibilities of modelling in biology.

In the quarter-century since, though, the model itself has only occasionally been re-used in plant developmental biology [28, 134]. One reason for this is the energy function $E(\cdot)$. While this is a mathematically natural formalism, it cannot easily be connected to any measurable phyllotactic process. In addition, while the model constraint that only one drop can be dropped at a time is crucial to its success, it is a limitation as a phyllotactic model because it forbids patterns in which two nodes are formed near simultaneously at the same height on the stem, although there is ample evidence that this happens biologically.

Fig. 9.2 Different patterns of ferrofluid drops in Douady and Couder's experiment at different values of the parameter G which control the rate of centrifugal movement. **a** When the time between drops is large compared to the time to be removed from the plate, each new drop is repelled only by the previous one and the node position alternates by 180° each time: in the language of this book a distichous pattern $(1, 1)$ is obtained with a divergence of $d = \frac{1}{2}$. **b** As the removal time decreases, a symmetry breaking bifurcation to a $(1, 2)$ pattern appears with a divergence $d \approx 5/12$. **c** As the control parameter is further decreased higher order Fibonacci modes, here $(5, 8)$, appear with divergence $d \approx 1/\tau^2$. ©Academic Press, 1996; from [33]

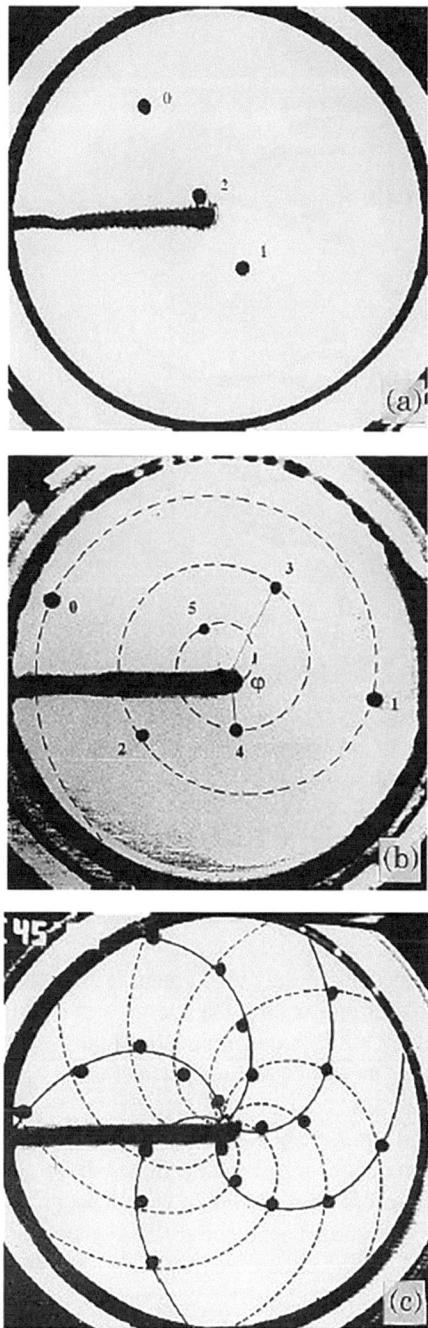

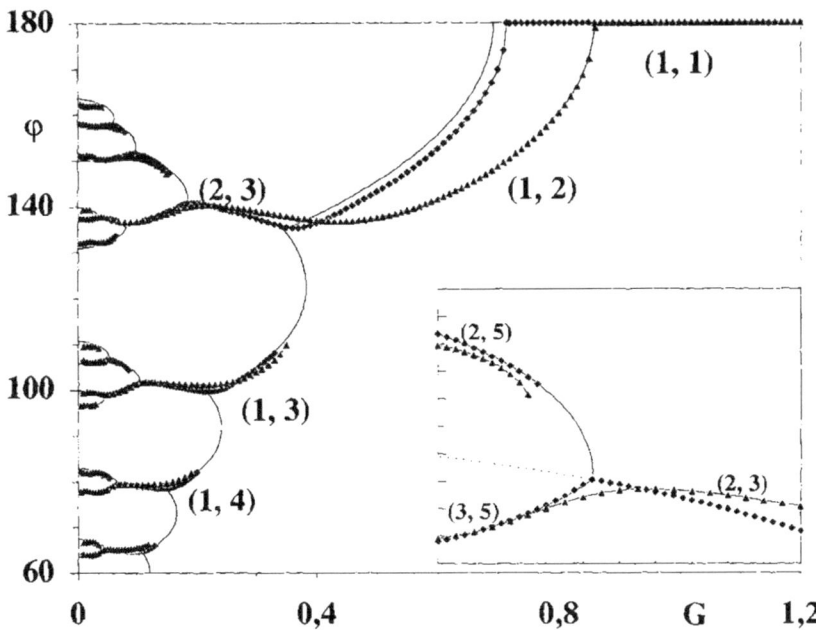

Fig. 9.3 Simulation results for a model of the Douady-Couder system, showing how only the Fibonacci branch is reachable with smooth changes of the control parameter G down from 1. The graph shows G, corresponding to relative magnitude of time to clear the plate to inter-drop period, against emergent divergence angle $\phi = 2\pi d$. From [33]

9.2 Energy Based Models

The Douady-Couder model of Sect. 9.1 imposed a time delay between new nodes and a centrifugal movement, which together ensured that new nodes could not be placed too closely to old ones. It then defined an energy function for the new pattern, so that minimising the energy step-by-step then placed nodes as close as possible given the first constraint. By contrast the stacked-coin model uses a fixed disk to keep the nodes separate, whilst using 'gravity', that is the requirement that the new node is in contact with the old ones, to keep them close. Both of these models yield a typical inter-node distance of say D, and can be thought of as defining a node-node potential energy which is large before D, then drops off sharply to a minimum near D before more smoothly re-increasing.

Given such a potential energy representing a mixture of close-range repulsion and long-range attraction, it is possible to define an energy for cylindrical node patterns, and then to analyse the patterns which are local minima of this energy. In the notation of Chap. 4, such an energy is of the form $E = \sum U(|\widehat{n} - \widehat{m}|)$ as a sum of functions of distances over node-pairs. A key difference between our application and most physical ones is that there is no mechanism for *global* minimisation of the energy:

a pattern may be built up node-by-node, locally minimising this interaction energy, but in general the final pattern is not a global minimum of all possible patterns on the same cylinder with the same mean D. Nevertheless it was these formal similarities with theoretical physics which led Levitov and others to the renormalisation technique [73]. If we restrict to patterns which are lattices, then the energy of a lattice can be analysed by renormalising the lattice, which simply scales all the node-pair distances by the same scalar. This analysis thus rediscovers the lattice bifurcation structure of the van Iterson diagram.

Early node-placement models such as those of Adler [2] were criticised by Lee and Levitov [69], I think unfairly, as not corresponding directly to any biological system. By contrast, Lee and Levitov claimed a deep universality for energy-based models as a justification for pruning the van Iterson tree to Fibonacci structure. However the energy function is defined in terms of 'distance', which has an intuitively reasonable interpretation in many field-based physical systems but is very hard to interpret here.

Nevertheless, this is a convenient formalism to express how the ratio of a fixed pattern length scale changes relative to developmental geometry. In a pioneering study in 1998, Yves Couder used such a model to show that it could generate some of the non-Fibonacci solutions described in Chap. 8 and seen in a sample, albeit a relatively small one of sunflower seedheads [28]. While these energy models have been relatively neglected in recent research relative to the stacked-disk models of the next chapter, their flexibility in expressing the idea of relatively soft disks and limited local rearrangements means they may well continue as a helpful tool.

9.3 Auxin Flow Models: Partial Differential Equation and Cell-Based Models

9.3.1 An Early Reaction-Diffusion Approach

In 1951 Alan Turing wrote to a prominent British zoologist of the time that

> …my mathematical theory of embryology…is yielding to treatment, and it will so far as I can see, give satisfactory explanations of
> (i) gastrulation
> (ii) polygonally symmetrical structures, e.g. starfish, flowers
> (iii) leaf arrangements, in particular the way the Fibonacci
> series (0, 1, 1, 2, 3, 5, 8, 13, . . .) comes to be involved
> (iv) colour patterns on some animals, e.g. stripes, spots and dappling
> (v) pattern on nearly spherical structures such as some Radiolara.[1]

By 'my theory of embryology' Turing meant primarily what we now call the Turing Instability, a pattern-formation mechanism for length-scale emergence in a reaction-diffusion PDE system. Turing's mechanism, which he later published in his cele-

---

[1] Letter from AM Turing to JZ Young 8 Feb 1951, King's College Cambridge AMT K.1.78.

brated 1952 paper [123] has been a mathematical blessing in its generality and a biological curse because of the same generality. With a few exceptions, biologists over the last 70 years have viewed the Turing instability as elegant but unhelpful or wrong for understanding the major questions in developmental biology. Subsequent mathematical biology using the Turing Instability has been particularly rich in (iv), is probably the wrong approach for (i) and (ii), and is largely forgotten for (v). But we are now in a position to see what Turing meant when he thought it could help with (iii). The previous models of this chapter abstracted known, or guessed, details of the molecular biology of node formation, into various representations of pattern formation on a cylinder with a typical length scale of D. It was the availability of such a model, in the form of the Turing Instability, that allowed Turing to develop *The Outline of the Development of the Daisy*. This was an unpublished manuscript in which he used a PDE model to explore transitions through lattice space and to try to find conditions for the Hypothesis of Geometrical Phyllotaxis to be fulfilled.

9.3.2 Cell-Explicit Auxin Flow Models

As molecular details have accumulated, Turing's essentially arbitrary activator-inhibitor model for primordia formation has had to be abandoned in its details. One modelling problem with any reaction-diffusion theory has been that auxin transport is not primarily diffusive, and is more importantly transported through the plant cell wall via PIN1-type proteins, in directions highly dependent on which faces of the cell wall PIN1 is localised. The Turing instability itself is unlikely to be a helpful concept for primordium formation. Its defining mathematical ability to generate a macroscopic length scale through a spatially extended linear wave instability seems not to play a biological role in these phyllotactic problems. Instead the length scale is generated, on our current biological understanding, by the physical scale of the developing primordium, which is largely independent of neighbouring primordia, and certainly distant ones, and controlled by a mixture of hierarchical genetic signalling, the hormonal and mechanical dynamics that maintain the primordium, and the need to fit against existing patterns.

PDE models, or their spatially discrete analogues which represent individual cells, continue to be a core tool in understanding the role of the spatio-temporal patterning of auxin in phyllotaxis, following [62, 95, 110], and have been particularly useful in articulating different hypotheses about the role of polarised auxin flux within cells. At present we do not have reliable enough quantitative measurements of auxin to fully validate the models at a cellular scale, and even recent work [87] validates the model through emergent node positions, just as we might with the stacked-coin model. However emerging molecular data at cellular and subcellular scales is likely to change this in the foreseeable future. Models such as those from the Smith laboratory [23, 78] are now impressively capable of generating detailed and testable hypotheses in conjunction with molecular work.

Some other researchers continue to claim simple continuum reaction-diffusion models as plausible source of phyllotactic pattern *de novo*. Carteni et al. [22] published a model suggesting that the patterning of vascular bundles could arise in this way, rather than as a sequence of patterns of increasing complexity. Among other problems, I doubt that such simple reaction-diffusion models can overcome the frequent problem of the strong dependence of emergent pattern on arena geometry.

9.4 Mechanical Tension Models

Mechanical stress on the rigid plant cell wall is known to reorient the microtubule orientation within the cell, and this has been suggested as a way in which PIN1 orientation, and thus auxin flux, can be affected by mechanical stress [54]. Galvan [43] is a recent review that includes this alternative tradition of modelling node placement as resulting from mechanical buckling, associated with Green [51]; recently models which include both biochemical and mechanical elements have been developed by a variety of different groups [19, 44, 83].

Indeed it is a significant advantage of a PDE-type formalism that it can be combined with a similar continuous stress-strain term using well-established formalisms. However when Pennybacker and Newell took advantage of this in their model, they still found they could derive all the key pattern behaviours they sought for Fibonacci structure by setting the strain dynamics to a constant [87]. But while the interaction between mechanical stress and molecular dynamics for phyllotaxis remains controversial, this class of models is likely to remain important [66].

9.5 A Divergence: Protractor Models

All of the models above are either consequences of embracing the Standard Picture or at least not inconsistent with it. An alternative explanation for Fibonacci phyllotaxis which has been greatly loved by mathematicians is that patterns are created by repeated rotation by the golden angle and that this has evolved because of a fitness arising from the resulting 'close' packing. The earliest appearance of this idea was Wright's 1859 suggestion that a uniform distribution of leaves would maximise sunlight capture [132], but despite being dismissed by high profile writers from Thompson [119] to Smith [113] it still persists; one goal of this book is to add to the chorus that these arguments really should be retired.

The most challenging modern voice to the Standard Picture is Okabe, who has used a series of peer-reviewed papers to argue that phyllotaxis shows 'exquisite control' of the divergence angle [85]. Okabe's papers are better sourced and more attentive to data than many proponents of the Standard Picture, but their reasoning is often hard to follow. Their prime claim seems to be that the Fibonacci angle is so finely on display in mature plants that it must have adaptive significance and have been arrived at by natural selection.

The circumstances under which an asymptotic approach to a uniform distribution over divergence angles creates a significant fitness advantage have never been demonstrated and it seems implausible they ever could be, given the sophisticated alternative shade avoidance mechanism available to plants. Close packing has also been claimed to prevent the ingress of pathogens, but this is more easily explained as a consequence of organs filling the Voronoi cells of their primordium patterns. As for maximising either uniform- or close-packing of the seeds of the sunflower, this could have some fitness, but it has never been identified. The sunflower has been subject to several thousands of generations, not of natural selection, but directed evolution by humans who have used it as a food source in the Americas. There must have been direct pressure to develop species with a large capitulum, but it is fanciful that any other morphogenetic factor played a role. But most fundamentally, the known biology offers no evidence for any molecular protractor in the plant delivering repeated divergence angles close to the golden angle and independently of previous primordia. Most current theorists would agree that precision in repeated instances of the divergence angle is, instead, an emergent property of pattern formation.

Okabe does make one observation that is useful to highlight here. The theorem that Jean christened the 'Fundamental' Theorem of Phyllotaxis says that a $(1, 2)$ system can take any divergence between $128.5°$ and $180°$, while experimental evidence is that divergence angles are in a much narrower range around the Fibonacci angle than this. Okabe's claim is perhaps that this suggests the divergence is a more fundamental property than the parastichy count, but my perspective is that this observation merely underlines how non-fundamental the Theorem actually is for Phyllotaxis: it classifies lattices, but does not describe how they have evolved from other patterns, not even nearby lattices.

9.6 L-System Models

L-systems, which use recursive string rewriting to encode increasingly complex morphological patterns, have long been used to generate phyllotactic patterns of great beauty [90, 92]. After the lattice theory, and especially the bifurcation theory, of this book, it is unsurprising that L-systems can generate Fibonacci-type phyllotaxis by placing the next node into the largest gap between existing ones. Recently these models have been put in much closer alignment with morphological and molecular evidence in some beautiful work on *Gerbera* [139]. Although to my mind the link between the biological dynamics of node placement and the emergence of this structure is less clear within the L-system framework, that is a matter of taste and these models will rightly continue to see considerable use.

9.7 Summary

This has been a non-exhaustive taster of range of models for node placement, and we have not yet reviewed one of the most promising contemporary models for node formation, the stacked-disk model which we will turn to in the next chapter. But 'promising' is a subjective judgement, and this book makes no claim to make final judgements on which models will turn out to be the most useful for future challenges. Readers looking for other perspectives can find them in [10], a modern book-length survey of placement models, with a rather different viewpoint from this one. Recent reviews mainly devoted to the biological issues include [121] and [46]. Reviews of mathematical modelling in plant morphogenesis generally are also worth seeking out [79, 81, 91, 92].

Chapter 10
Stacked-Disk Models

Abstract Disk-stacking models for stem development have recently emerged as an excellent compromise between mathematical simplicity and biological relevance. Because disk-stacking models allow lattice solutions, the van Iterson paradigm remains a powerful organising principle for their dynamics, and it is by now well established that disk-stacking models can indeed demonstrate Fibonacci structure (Golé et al. in Acta Soc Bot Pol 85(4), 2016). Moreover, as the van Iterson paradigm suggests, there are parameter regions in which strict Fibonacci patterns are lost but patterns remain closely ordered with either Lucas numbers or double-Fibonacci pair counts occurring (Golé et al.; Yonekura et al., PLOS Comput Biol 15(6), 2019). Disk-stacking models have now been shown to exhibit further empirical phenomena in ways no other models have yet achieved or seem likely to (Swinton, arXiv: 2407.05857). However there remain many open questions, both mathematical and biological about the dynamics of this interesting class of models.

10.1 Motivation

A stacked-disk model represents a series of disks, placed vertically in sequence around the outside circumference of a cylinder, in such a way that each disk is placed at the lowest point it can be without overlapping any of the other disks (Fig. 10.1). The disks of the model represent the inhibition zones in the apical meristem that are established by patterning of auxins and other morphogens, and the 'gravity constraint' of adding the new disk at the lowest possible place corresponds to Snow's empirical rule that a new node will form as soon as it has enough space. The changing disk size corresponds to the change in proportion between the sizes of the inhibition zone and that of the apical meristem diameter. It seems likely that most of this change in relative geometry in actual plant growth is due to expansion of the apical meristem rather than modification of the inhibition zone size. Figure 10.2 gives an example of a model in which the inhibition disk size is fixed at the size of a British 2p piece, and the apical meristem is assumed to increase linearly with height, and periodicity of the pattern is judged by eye. This book uses the terms 'stacked-disk' and 'stacked-coin' interchangeably.

Fig. 10.1 Schwendener's stacked-coin model showing transitions from a (3,5) to a (5,8) to a (8,13) parastichy [106]

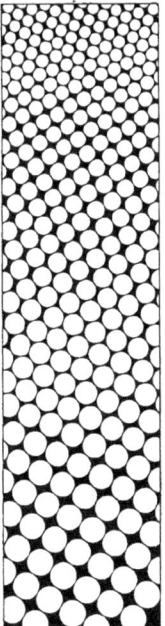

A closely similar model, and a more computationally convenient one, is instead to scale dimensions so that the apical meristem circumference is fixed (and moreover fixed at unity) but the inhibition disk sizes vary as in Fig. 10.3. The two model classes are not quite identical because the map between cone and cylinder will distort circles, but to the extent that the results of model simulations are not strongly dependent on the shape of the inhibitor zone, we can expect similar results between the two. We can think of the existing disks creating an inhibition field, shown in grey in the figure: no new disk can be placed below the boundary of the region formed by disks centred on all the existing disks. The new disk is placed at the minimum of this boundary. Attractively, this placement depends only on the most recent 'chain' or 'front' of disks around the cylinder, and is the minimum of a finite number of intersections of disks.

10.2 Stacked-Disk Models

The primary modelling choice is how to choose the inhibition diameter D of each disk relative to that of the cylinder. Although not in principle completely out of empirical reach, precise quantification of this parameter, and how it varies over the course of plant development is in practical terms not feasible at present. For modelling purposes we will be content with rough estimates in the order of magnitude changes

10.2 Stacked-Disk Models

Fig. 10.2 A stacked-disk model, implemented with British tuppeny pieces, inspired by Atela [6]

Fig. 10.3 Finding the next location in the stacked-disk model by finding the lowest point of the boundary of the inhibition zone, itself defined by the topmost coins. The width of the inhibition zone may be variable and affects the position of the next disk. Left, the lowest point not in any inhibition zone is marked with a red dot, and (right) the largest possible circle drawn around it

to be expected from relative changes in SAM geometry relative to the spatial scale of the molecular dynamics. All of this biology is then encoded in a function $D(z)$ where z is the height so far achieved up the cylinder.

Supposing that we have placed disks at heights z_1 up to z_k, there is a subtle implementation point about which z should be chosen to decide the radius of the next disk. One coherent approach is to say a new disk centred at z_{k+1} should have inhibition diameter $D(z_{k+1})$. However, finding the lowest possible next such disk then requires a non-linear minimisation at each step. A more computationally expedient approach, adopted in all the simulations in this book, is to set the inhibition diameter of the $k+1$th disk to $D(z_k)$; then the next disk position is found as the minimum of a finite number of arc-endpoints. Given the inherent simplifications involved in representing the molecular biology of inhibition as a single continuous function $D(z)$ it is unlikely empirical evidence can be used to prefer one approach over the other; in any case two approaches seem unlikely to cause any significant major differences in results, although this has not been explored.

In general we will be interested in $D(k) = D(z_k)$ as a decreasing function of the height z_k of the most recently placed disk, although when we look at the generation of sunflower capitulum patterns D will be first decreasing until the capitulum rim is reached and then increasing as seeds are placed towards the centre.

10.2.1 Stacked-Disk Models Cannot Generate Non-Opposed Lattices

If we take one of the touching-circle lattice patterns of the first half of this book, with a disk diameter D, and truncate it above one point, then we can use the disk-stacking model with that fixed D to regenerate a pattern. As Fig. 10.4 shows, if the original lattice was an opposed lattice then the first iteration of the model will exactly replace the disk where the lattice had put it, but if it was a non-opposed lattice then it will not do so.

The reason for the latter is that the placement mechanism guarantees the placement of a disk of a given radius at its lowest possible point, and the contact lines from a newly placed disk down onto its two supports must be in opposite directions on the cylinder. If they are not the newly placed disk could be 'rolled' down to a lower placement. But in a non-opposed lattice, by definition these contact lines go in the same direction on the cylinder.

10.2.2 Parastichy Numbers in Stacked-Disk Models

While stacked-disk models *can* generate lattices, usually they do not. So we can no longer rely on the definitions of Chap. 4 to find a global pair of parastichy numbers for the pattern, because parastichies are not exactly straight lines. However, they are not so far off, and as Golé and Douady pointed out there is still a way to compute

10.2 Stacked-Disk Models

Fig. 10.4 Non-opposed lattices cannot in general be generated by stacked-disk models. Given the white disks as an initial condition formed from a subset of the non-opposed lattice of Fig. 5.3, a next disk placed according to the lattice would not be the lowest possible

parastichy counts locally [48]. The process is shown in Fig. 10.5. Starting from one disk, we look for a chain of touching disks that encircle the cylinder and return to the original. That chain will have a number of up-steps to higher disks and down-steps to lower disks, and the total number of up-steps and down-steps is exactly what we define as the parastichy count pair. There is in general more than one way to find such a chain through any particular disk. If the pattern was one of the touching-circle lattices of Chap. 4 this would coincide with the parastichy counts there no matter which chain was chosen. But when triangles are present in the pattern, as in Fig. 10.5, the precise choice of path can modify the counts. One might choose the shortest possible chain, or as in the highlighted chain of Fig. 10.5 the highest chain through all the disks lower than the initial disk. These different choices affect exactly where on the cylinder the parastichy numbers appear to change, but not what those ultimate changes are, and the simulations in this book use the highest chain method.

It is in this way that the discrete-valued parastichy counts can transition by increases of 1 as the disk radius reduces slowly. Each of these jumps corresponds to a 'γ' dislocation (taking its name from the shape of the letter [136]) of the parastichy lines; conversely, λ dislocations correspond to a decrease in the count. This method coincides well with human assessment of spiral counts in relatively well-ordered patterns but can still be assigned in strongly disordered patterns when the human eye is unable to identify structure.

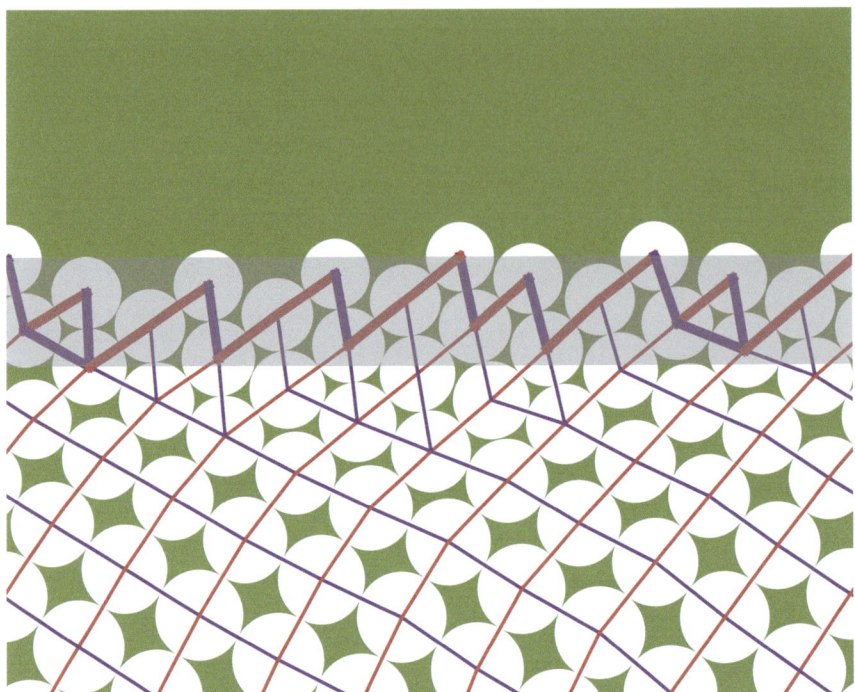

Fig. 10.5 Computing the parastichy counts for non-lattice patterns by counting up and down links in chains around the cylinder. If the patterns is close to a lattice, these counts coincide with the parastichy numbers of the lattice. In this example, there are 8 red parastichy lines going upwards (when looking left to right); correspondingly the topmost chain of connections between disks has 8 separate segments of sequences of upwards lines. At the bottom of the figure there are five blue segments of downwards lines corresponding to a 5-parastichy, higher up there are 8 such segments, reflecting a series of increases in the left parastichy number from 5 through 6 and 7 to 8

10.2.3 Stacked-Disk Models Naturally Generate Fibonacci Transitions

The principal result of Chap. 5 was that if we had a mechanism, as stacked-disk models do, that generated lattices, but rejected non-opposed lattices in favour of opposed ones, then Fibonacci structure could result. Indeed solutions to stacked-disk models showing large Fibonacci pair parastichy counts were exhibited by Bursill and colleagues in the 1980s [21] and more recently and systematically investigated by Golé et al. [48]. As an example, Fig. 10.6 shows a (5,8) to (8,13) transition, and in general numerical observation supports the principle that patterns with parastichy pairs (m, n) with $m < n$ do indeed typically transition to ones with $(n, n + m)$ parastichy pairs provided that the dimensionless number $D' = dD(z)/dz$ is small enough. It would be surprising if this was not in general true, at least near Fibonacci structure, but there is currently no analytic proof of this. An understanding of the van Iterson

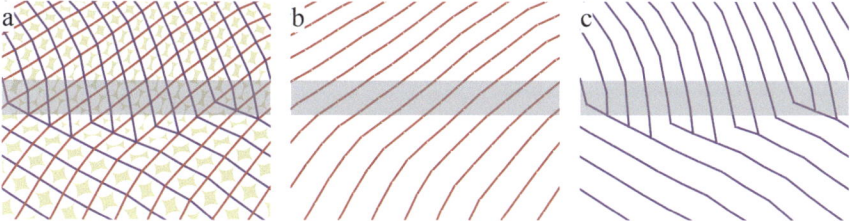

Fig. 10.6 Transitions between a (5,8) parastichy pair and an (8,13) pair as disk radius is slowly decreased. At the bottom of the figure the disk pattern is close to a square (5,8) lattice. During the transition region the disk arrangement is close to a hexagonal lattice, and then after the transition the internal lattice angle widens from 60° as the pattern becomes close to a (8,13) square lattice

tree adds further insight to these smooth, deterministic transitions. In van Iterson's parameter space, lattices change their parastichy numbers exactly when they are hexagonal lattices, then widen in internal angle into square lattices before narrowing again into hexagonal lattices at the next lower transition. For slow rates of disk radius change, the disk pattern at the transition is very close to a hexagonal lattice, and the transition is able to occur smoothly. In particular, the positions of lattice points above and below the transition are strongly correlated: the change of parastichy number occurs by a smooth changeover between which of these points are closest to each other.

So in general, it seems to be the case that for small enough D', stacked-disk models satisfy Turing's hypothesis of geometrical phyllotaxis. To generate Fibonacci, rather than say Lucas numbers, we need to look at the ways in which the (1, 1) patterns corresponding to stacked disks of diameter $D = 1$ change as D is slightly decreased, and it is not difficult to see that for small enough D' the next parastichy pair must be (1, 2). In general, then, there are strong intuitive and numerical grounds for expecting that stacked-disk models with small D', starting from $D = 1$, will always generate Fibonacci parastichy pairs, as indeed occurs in Fig. 10.7. So this disk-stacking model can account for the dominance of Fibonacci counts in the empirical data of Fig. 8.3. What it does not by itself account for is the other peaks in that data: we explore that later in this chapter.

10.3 Outstanding Questions

It is tempting to stop at this point: we have built a model of node formation which is informed by and fairly consistent with the known molecular biology, and we have shown that as a biologically relevant parameter is varied, the model outputs pass through a series of increasingly complex patterns, each transition preserving the Fibonacci property. More than that, this property is generic and does not rely on model fitting. Apart from ensuring that the change of disk radius is slow enough, we

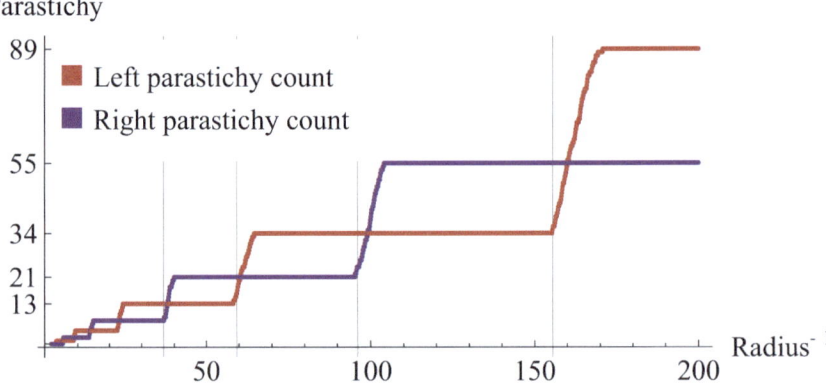

Fig. 10.7 Transitions between parastichy counts as a function of the slowly changing inverse radius of the disks. y axis: Local chain parastichy counts for the topmost chain through each successive disk in a single deterministic run with a $r' = 0.03$ started from a single disk of radius 1/2. x axis: inverse radius of the disk at the highest point of that chain. The thin grey vertical lines at $r^{-1} = 2\sqrt{2}F_k$ mark the start of transition from F_{k-2} to F_k

have had to specify no particular parameters to achieve this sequence of Fibonacci transitions. However there is more going on in the dynamics of these systems than transitions between lattices, and this more mathematically complex dynamics appears also to have at least some biological relevance. So showing the relevance of the van Iterson classification requires overcoming a (to me) surprising mathematical obstacle: when is a lattice adequate to describe stacked-disk dynamics?

10.4 Coherent Structures in Stacked-Disk Models

Does a stacked-disk model reliably lead to lattices? Intriguingly, the answer is no. Figure 10.8 shows the results of five simulations of a fixed-size stacked-coin model, of which the centre starts from an exact (2,3) lattice, and the neighbours start from a perturbation of that lattice. On the left we can see our square (2,3) lattice with straight line parastichies. For the other initial conditions with the same disk radius the lines joining the disk centres are now longer exactly straight, yet would still be described by eye as having 2-parastichies and 3-parastichies, although not always in the same direction.

So a stacked-coin model can lead to solutions which are close to lattices, and if we start with a disk radius corresponding to a (m,n) lattice and an initial condition close to one, we might hope to remain with a (m,n) parastichy count in our generalised sense. We can also hope that, just as the new third parastichy number at a lattice bifurcation is either (n+m) or (n-m), the same will be true for patterns near to those lattices, and indeed Golé et al. [48] give an elegant argument based on chain transitions for this.

10.4 Coherent Structures in Stacked-Disk Models

Fig. 10.8 A stacked-coin model with disk size fixed at that of a square (2,3) lattice, started from different initial conditions in which a single disk of the square lattice has been distorted in size

But how close to a lattice do these patterns need to be for the elegant bifurcation theory of Chap. 5 still to hold?

10.4.1 Cylinder Tilings and Rhombic Tilings

One way to quantify the way in which the runs of Fig. 10.8 depart from lattice structure is to plot a function associated with each node, such as the area of the polygon below it, as a function of the dropped-coin sequence. This is done in Fig. 10.9. These seem to be periodic graphs, corresponding to periodic orbits of the stacked-coin map.

Just as a lattice corresponds to a fixed point of the next-node vector stacked-coin map, with a constant (d, h) between nodes, a cylinder tiling corresponds to a low-order periodic solution of the map, with a sequence $((d_1, h_1), \ldots (d_k, h_k))$. The upper parts of the tilings in Fig. 10.8 are close approximations to cylindrical tilings. A consequence of the disk-stacking model is that each node in Fig. 10.8 is generically at the apex of a single polygon. (Exceptionally, if the pattern is a hexagonal lattice, there will be two such polygons.) More than that, in this figure at least, all of the polygons of the tiling being approached appear to be rhombi: four-sided polygons with opposite parallel sides. A cylindrical tiling in which the polygon attached to

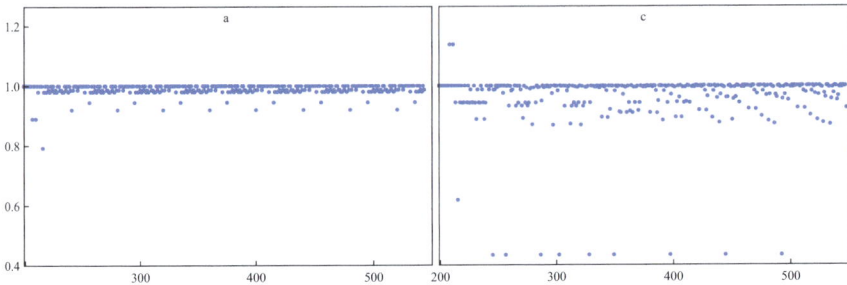

Fig. 10.9 Area of each polygon in the lattices of Fig. 10.8, read left to right, plotted from bottom to top

each point is one of a finite number of different rhombi was called a rhombic tiling by Atela and Gloé [7] who introduced them in this context.

Figure 10.10 shows an example in which the stacked-coin map is visually close to having periodicity of 6. The map trajectory defines a series of polygons attached to the coin centre which is its lowest point. In this rhombic-tiling pattern each coin has exactly one rhombus attached to it.

Douady and Golé gave an interesting example of a cylinder tiling and a lattice separately fitted to the same observation, which is shown in Fig. 10.11. Since a tiling made up of n different rhombi has at least $2n$ degrees of freedom, it can naturally give an better fit to an observed pattern than a lattice, which has $n = 1$, and it is an open question whether empirical observations can at present reject lattice patterns in the framework of statistical hypothesis testing.

It is striking that before these quasi-lattice patterns were seen in model outputs, there were few attempts to parameterise empirical data in any way other than as realisations of a lattice. One notable exception was Atela et al. [8] who suggested that a period 8 orbit they found in a placement map corresponded to a similar periodicity in divergences observed on a magnolia carpel [122]. Beyond this example, shifting the unit of analysis to a cylindrical or rhombic tiling has generated an interesting new series of observational questions [35].

Rhombic tilings are also important mathematically because, as Fig. 10.8 illustrates, small perturbations to initial lattice patterns typically evolve into rhombic tilings under fixed-size coin stacking [7]. Golé and Douady have proved that every initial chain that has a parastichy count of $(1, 2)$ will indeed evolve into a rhombic tiling in finite time or become exponentially close to one, and conjectured this is true in general [47].

It has been suggested that stacked-coin disk provides an explanation of a 'butterfly effect' in which the phyllotaxis of, e.g. the magnolia stem sometimes changes 'for no obvious reason' [137]. Numerical simulations can indeed sometimes show surprisingly long intervals of apparent equilibrium followed by a burst of changes, and even if these trajectories are only transient and may be of value in exploring phyllotactic transitions in species like *Magnolia*.

Fig. 10.10 Rhombic tilings arising from a stacked-coin model. Tiles are coloured to emphasise a periodicity of 4

10.5 Columnar Models

Exploration of the dynamics of stacked-coin models has revealed another intriguing pattern-formation process. If instead of very slowly changing the coin radius we very quickly reduce it and then keep it small, we are effectively starting a fixed-radius run with random initial conditions and relatively small disks compared to the cylinder circumference. Figure 10.12 gives an example. It turns out, for reasons explored in [48], that runs often converge to a series of near horizontal sets of disk stacked on top of each other. This stacking is fairly regular so that disks in either every row or every other row are vertically above each other. In the older language of phyllotaxis, these patterns might be called whorled orthostatic. Golé and Douady [47], who

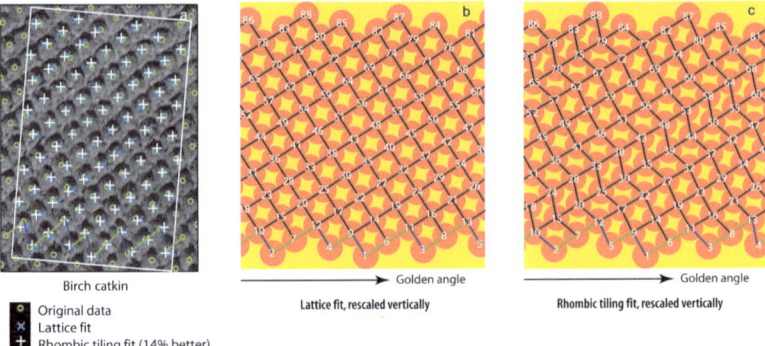

Fig. 10.11 Fit of a lattice and a cylinder tiling to the same unrolled birch catkin image. The cylinder tiling is a 14% better fit to the node positions than the lattice. From [35]. © CC-BY Douady and Golé 2016

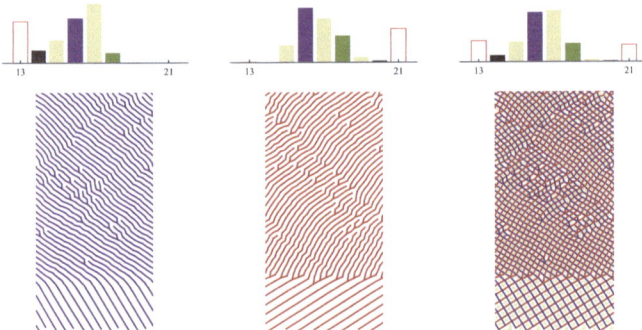

Fig. 10.12 Runs of a stacked-coin model for increasingly steep radius changes. In each case the disk radius is reduced from that of a (2,3) orthogonal lattice to that of a (5,8) one. When the transition is slow enough for an intermediate (3,5) lattice to establish well before moving into the zone where a (5,8) lattice is possible, that transition is visible is followed by an equally ordered transition to what appears to be a rhombic tiling close to a (5,8) lattice. When the disk radius is reduced more rapidly, the ordered transitions are lost. In the central run, the parastichy count at the top of the cylinder is (7,8) while in the right-hand run with a still faster reduction of radius the parastichy count at the top is (7, 7)

have done unparalleled work on them, call these QSS patterns, for quasi-symmetric solutions, but here I call them *columnar* patterns.

This also provides a potential pattern mechanism for columnar patterns like those of the sweetcorn of Fig. 8.10 or of cacti. These models might well replace Turing Instability-based ones as a plausible pattern-formation mechanism for such patterns.

It is clear numerically that the elegance of pattern analysis as transitions through near-to-touching-circle lattices is lost if the geometric change is too fast, and this can be interpreted as being because the close-packing property is lost: the constraints of a lattice pattern cannot adapt quickly enough to stay close to being well packed, and lattices as the units of analysis become unhelpful. But the extra degrees of freedom of rhombic tilings might well allow them to adapt more efficiently, in terms of close-packing, to geometric change than lattices do. So it might be that Fibonacci transitions through rhombic-tiling space can be maintained at higher rates of geometric change than if lattice structure were enforced. Arguments have already been made that rhombic tilings have their own more general version of the Hypothesis of Geometrical Phyllotaxis, and so taking the rhombic tiling as the unit of analysis rather than the lattice might provide a more biologically robust explanation of Fibonacci phyllotaxis than one based solely on lattices.

10.6 Finite-Time Dynamics and Sunflower Data

So stacked-disk models provide an attractive generalisation of lattice models for exploring Fibonacci phyllotaxis, although their mathematical properties are much well less worked out, even for fixed disk size. For example, it is unclear if every attractor of such models is a cylinder tiling, or under what circumstances such models can yield disk packings which are denser than those of the corresponding lattice. Despite these uncertainties about the asymptotic dynamics of the models, they are still extremely useful as a lightly parametrised model that is capable of explaining more than the Fibonacci structure in comparison with data.

To do this we need some data to compare with, and as we have seen all of the possible datasets so far available have their limitations. Couder's 1998 dataset of 21 sunflowers is notable for not only containing, by design, both Fibonacci and non-Fibonacci specimens, but for recording both leaf and seed patterns on the same specimen [28]. The largest current dataset, though, is the MOSI dataset discussed in Chap. 8. We recall the following features:

1. There was a strong but not complete preponderance of Fibonacci counts.
2. Excluding Fibonacci counts, the next commonest parastichy count was one less than a Fibonacci number (specifically, 33, 54, or 88). A Fibonacci number less one (like 33) was statistically significantly more likely to occur than a Fibonacci number plus one (like 35).
3. The next most common was a Lucas number (29, 47, or 76), and then a double-Fibonacci number (42, 68).
4. It was common to see sunflower heads in which parastichy spirals could be clearly counted in one direction but not in another.
5. In a small number of relatively small sunflower heads, pairs of nearly matching but non-Fibonacci parastichy counts like (11,11) were seen.

Bearing these criteria in mind, Fig. 10.13 gives an example of multiple replications of a stacked-disk model with a stochastic element, described further in [114]. While the match between the empirical data and the simulations is far from exact, these models, alongside Couder's pioneer work [28] are the first published ones capable

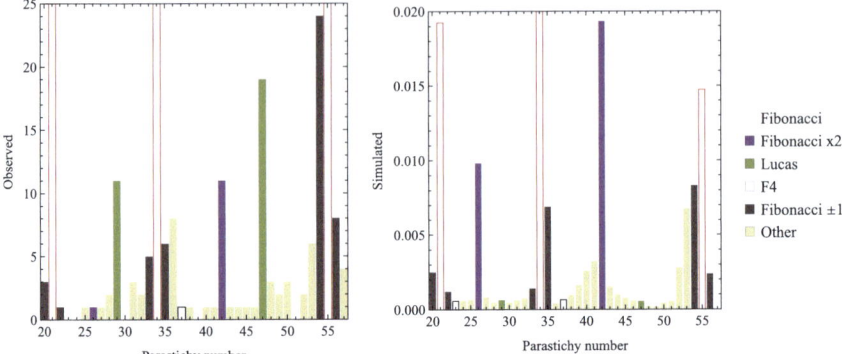

Fig. 10.13 Left, empirical parastichy counts redrawn from Fig. 8.3 to emphasise non-Fibonacci observations. Right, parastichy counts in a disk-stacking model with $r' = 0.03$ and noise $\sigma = 0.05$. In each simulation the initial condition was a single disk of radius $r(0) = 1/2$, with a disk size function $r(z) = 1/2 - r'z$, run until $r(z)$ changed by a factor of 60, corresponding in a lattice model to a change from a (0,1) lattice to a (34,55) one. Simulated parastichy numbers were pooled over the course of each run and then further pooled over 10 replicates of the randomisation. Redrawn from [114]

of generating Lucas and double-Fibonacci numbers as well as the occasional non-Fibonacci structure in this way.

10.6.1 Cylinder to Capitulum Mappings

The output from disk-stacking models can be mapped from cylinders to any surface of revolution, as in Fig. 10.14 which shows how results of disk-stacking models can be compared with empirical data on spirals at the outer sunflower seedhead rim.

Here we have first used a disk-stacking model to simulate a rising and then falling phyllotaxis modelled using a radius function whose inverse first linearly increases then linearly decreases. Only after this have we, arbitrarily, added functional labels and corresponding colours to each of the placed disks. After nodes labelled as uncommitted, bracts, and ray florets, and above an arbitrarily chosen point z_S, nodes are deemed to correspond to seed positions in the mature seedhead. Still higher node positions are deemed to correspond to a disk floret which does not proceed to set seed, as is common in the centre of large seedheads, up to the end of the run at $z = z_U$. Right parastichy lines are shown in the region which will correspond to mature seeds in the adult seedhead. Below: the resulting placement pattern and parastichy lines mapped onto a seedhead disk. Points at coordinates (x, z) on the cylinder are mapped to radial coordinates $\rho = (z_U - z)/(z_U - z_S)$, $\theta = 2\pi x$. Note that occasionally some disks which are in contact in the cylinder pattern, for example, in the ray florets), corre-

10.6 Finite-Time Dynamics and Sunflower Data

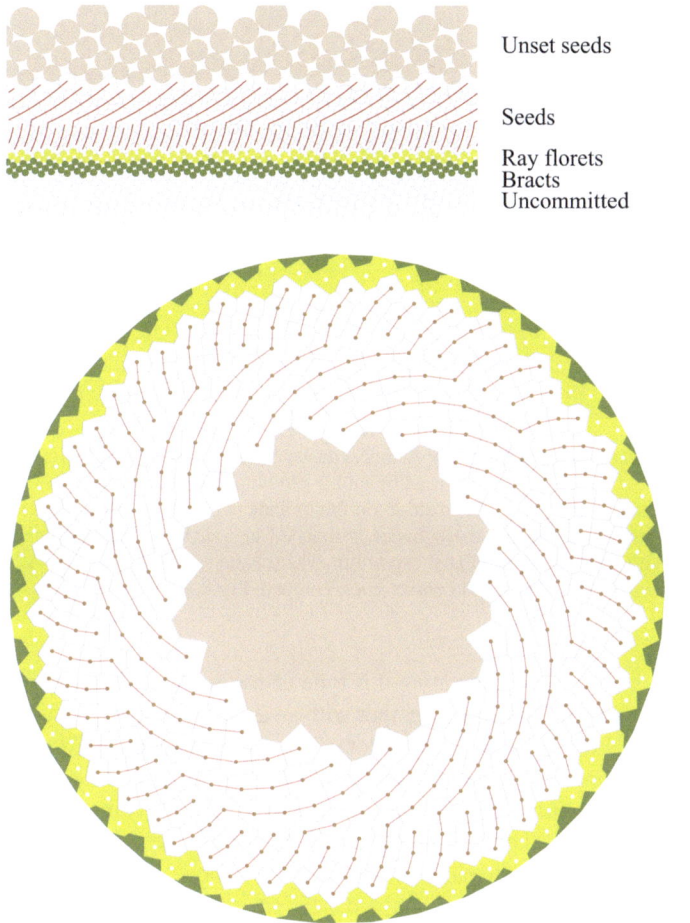

Fig. 10.14 Mapping a disk-stacked pattern to a seedhead pattern as described in the text

spond to Voronoi cells which are not in contact, partly because of the non-isometry of the mapping from cylinder to seedhead disk.

10.6.2 Falling Phyllotaxis and Observed Asymmetry on the Capitulum

We saw in Chap. 8 that the 2016 MOSI dataset exhibited a previously unreported departure from Fibonacci structure, in occasional specimens with a lack of rotational symmetry in the assignable parastichy lines [116]. The stacked-disk model for falling

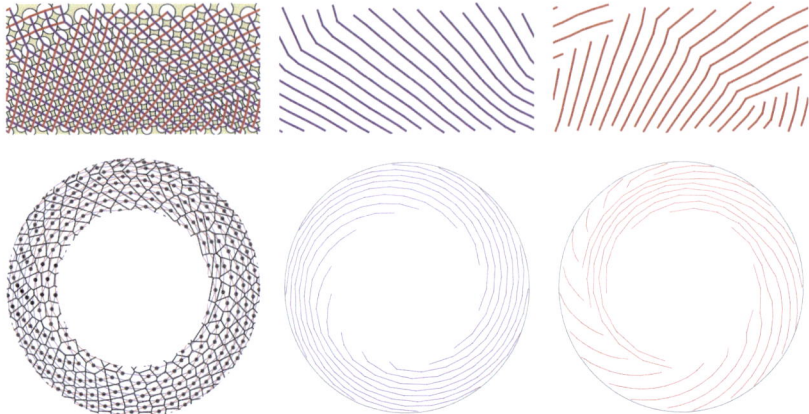

Fig. 10.15 Stacked-disk models on the cylinder can generate radially asymmetric patterns by rising phyllotaxis to the capitulum rim followed by falling phyllotaxis on the capitulum. This simulation shows horizontal variability in the location on the cylinder of dislocations of the red right-ward leaning parastichy lines. When the cylinder is mapped to a disk, this corresponds to a visible rotational asymmetry on the simulated capitulum. There is no satisfying way of assigning 'spiral counts' on the outer rim for the red-contact lines: compare Fig. 8.6

phyllotaxis is capable of illustrating this type of pattern, as illustrated in Fig. 10.15, which shows such a simulation, together with an arbitrary mapping of the simulation back to the capitulum. For rising phyllotaxis, the transition zones where parastichy number change are very narrow and horizontal, even in the highly disordered example of Fig. 10.12. By contrast, the transition zone in this falling phyllotaxis example is not horizontal, but follows the gradient of the parastichy lines and so there is a large region of the cylinder on which the parastichy counts are difficult and perhaps meaningless to assign. Looking at the lower right panels we can see how the uncountability manifests itself: there is a family of 21-ish red-contact lines making one angle in the cylinder, and another family of 8 red-contact lines making another, but the two families coexist over an extended region of the cylinder. This is the pattern that was identified in sample 667 of the MOSI image, whose hand-assigned parastichy lines are reproduced in Fig. 8.6.

10.7 Notes to this Chapter

Disk-stacking models were first introduced by Schwendener in the 1870s as an exploration of organ placement in plants [107]. Schwendener sketched the patterns seen when disks of decreasing size were stacked, one after the other, around a cylinder (Fig. 10.1). Although Schwendener's work was not influential on mainstream plant morphology, it seems to have influenced the better known 1907 PhD thesis of van

10.7 Notes to this Chapter

Iterson. Van Iterson who explored the possible patterns arising from arrangements of fixed-size disks as a function of disk size and first drew a version of the van Iterson diagram [127]. A version of this model in which the fronts were curves called 'pseudoconchoids' was arrived at by Schoute at the first quarter of the twentieth century [103]. According to Richards' account [97] of this German-language work, a pseudoconchoid allowed phyllotactic transitions by giving the front shape 'sufficient flexibility'. But the attention of mathematicians in the twentieth century was almost entirely restricted to elaborating the van Iterson lattice theory described in the first half of this book. It was not until the 1990s that van der Linden demonstrated the first simulations of a Schwendener-type model showing large Fibonacci numbers [126].

The twenty-first century has seen a revival of interest in the dynamics of these stacked-disk models [2, 6–8, 46, 56]; this book has been particularly strongly influenced by [48] and [47].

Part IV
Conclusions

Chapter 11
Some Future Directions

Abstract Although it is the goal of this book to equip readers to decide future research priorities for themselves, here are some brief observations on the kinds of mathematical investigations that could contribute to a mathematical biology of phyllotaxis. These cover both what are conventionally called pure and applied mathematical topics.

11.1 Mathematical Issues

11.1.1 The Van Iterson Diagram

The van Iterson diagram is now of venerable age, and is normally analysed for monojugate lattices with $J = 1$, although many writers have recognised that multijugate lattices with any fixed $J > 1$ can also be described by a copy of the van Iterson lattice. These slices of lattice space can be glued together at the $h = 0$ boundary, and this provides a path in lattice space along which parameters can be smoothly changed, but the examples of the stacked-coin model show that transitions between monojugate and multijugate lattices, which are commonly seen, are not transitions of this form. It would be valuable to re-visualise lattice space in ways that make these transitions seem more natural. The renormalisation transform that structures lattice space can itself be used to map between monojugate and multijugate lattices, and this may provide a tool for this.

11.1.2 Cylindrical Tilings

The stacked-coin model is particularly useful for drawing attention to cylindrical tilings rather than lattices as a basic unit of pattern. Relative to planar polygonal tilings, which have been widely studied (and yet only yielded the surprise of the Penrose tiling in the 1980s), cylindrical tilings have been less intensively analysed, and again might be worth considering in the light of a tiling renormalisation that exploits the cylindrical periodicity. The stacked-coin model itself possesses mathematical structure, in the form of conserved quantities and the periodicity constraint, that are unlikely to have been fully worked out and may well yield a cornucopia of dynamical systems behaviour beyond the interesting cases already known.

11.1.3 Regular Patterns as Attractors

The central prize from more mathematical work on stacked-coin or other models would be to get a good general understanding of when and why lattice-like structures might be expected to emerge from the dynamical systems of phyllotaxis at all. For example, is it really the case that it is the cytokinin feedback that stabilises the dynamics of node-formation order, as [13] suggest, or is this an inevitable consequence of any evolvable regular pattern-formation process? We do not have an analytical expression for how slow geometrical change needs to be to preserve lattice structures nor do we have good understandings of whether different lattices (e.g. Fibonacci) are more stable than others under such changes.

It is likely that other models, perhaps with softer-edged coins or other longer range interactions, will smooth over the difficulty the hard stacked-coin model has in converging to lattice patterns, while still providing a simple connection, via the Hypothesis of Geometrical Phyllotaxis, with the appearance of Fibonacci structure.

11.2 Modelling Challenges

11.2.1 The Phyllotactic Jump

One missing link in the Standard Picture is the observed leap, in the *Compositae* at least, between parastichy counts in the stem and in the disk floret. The common daisy has a handful of leaves, all branching at ground level, whose parastichy structure has never been systematically reported, then a stem which is completely smooth to the naked eye, before the bract, ray floret and disk floret structure appears. That the daisy usually has 13 bracts suggests a transition through 3 and 5 and 8 in the principal parastichies of the node pattern during development: at a minimum that requires 8 nodes, and probably rather more. But where are they? The situation is even more dramatic in the exaggerated seed disk of the sunflower. Sunflower stems have been the object of phyllotactic study, but typically report $(1, 2)$ or $(2, 3)$ parastichy counts if any at all can be evaluated. As we have seen, bract counts in the sunflower are not tightly clustered on Fibonacci numbers, though the Standard Picture can explain this away by saying that the bract generation region is not necessarily a tight annulus representing a single front. But even so the Standard Picture does demand that node placement in the bract region relies on a pattern predisposition with at least 13 nodes

11.2 Modelling Challenges

and probably more. Faced with the smooth underside of the disk capitulum, there is no visual evidence of the past trace of such a pattern. It is perhaps no coincidence that the major modern proponent of the divergence angle as the driver of pattern, Okabe, has paid particular attention to the low-order pre-patterns that are in practice seen [84]. Of course it is common in development for early commitments to be erased later, but whether this has left more subtle anatomical traces than are visible, or whether the right molecular probe at the right stage of development can yield the missing intermediate parastichy counts, remains to be seen. Relatedly, the Standard Picture naturally explains Lucas, or double-Fibonacci patterns as evolving developmentally from an initial (1,3) or (2,2) pattern say, and Couder's insightful study linking stem- and seedhead-phyllotaxis provides some limited support for this [28].

At a molecular level there is as yet only tantalisingly sparse experimental evidence about where this switching between branches actually occurs in development. Such evidence will be very experimentally demanding, but it is already the case that molecular probes visualising genetic or protein activity at cellular scales, stacked using confocal microscopy across the whole geometry of the shoot apical meristem, are providing whole new arenas in which models need to be developed and tested.

11.2.2 Macroscopic Tests of the Standard Picture

As we have seen, older theoretical views of stem form often envisaged the divergence as the central controlling parameter. In that viewpoint, the sequence of organ formation is essentially implicit: the 'next' organ is the one placed at the next rotation by the divergence. Correspondingly, changes in global phyllotactic counts have to be interpreted as changes in the order of node production, as in the otherwise sophisticated statistical analyses of [52]. However the Standard Picture, as presented in this book, rejects the idea of the divergence as a morphological parameter under direct genetic control, and correspondingly allows observed macroscopic pattern changes to be interpreted as the consequence of local node-node interaction dynamics.

Even macroscopically, it is feasible to test the Standard Picture. Is the distribution of ordered but not strict Fibonacci patterns of the types enumerated in the MOSI experiment consistent in other studies? Is it consistent with deviation from Fibonacci seen in the fir cone studies? And are these distribution modellable as the outcomes of, say, stacked-coin models which differ only in the speed of geometric change? Columnar lattices, though a-priori rather obvious as good packings of relatively small disks on cylinders, were never looked for empirically until after modellers like Douady and Golé pointed out their existence as a class of model solutions. It is rather instructive that before there was a concept of such lattices it was very hard to see them, and now they are easily detectable.

Beyond this, can the additional fitting power of cylindrical tilings that Doady and Golé found [35] be shown to be statistically significant beyond the additional degrees of freedom, and is it feasible to use these as the basic unit of empirical observation? While recent work using discrete Fourier transforms to quantify phyllotactic patterns is valuable [82], these tools need to be widened to consider cylindrical tiling patterns. Anyone who has analysed these patterns will yearn for a good image-analytic toolkit to take the drudgery out of finding the coordinates of nodes and there has been some promising recent work on this [3, 4]. As three-dimensional scanned representations of objects such as fir cones become increasingly feasible at scale a new class of macroscopic analysis in three dimensions will become possible.

When untangling the interplay of growth and development in the mature plant one useful principle is the near-universal lack of observations of organs significantly changing their angular position on the stem once committed, and the frequent observation of golden angle divergences in mature plants is a core constraint on any model of what is going on in the developmental stage. Some writers have gone further than this and argued that the developmental processes themselves must be controlled by divergence-based mechanisms, but it is the view of this book that this is quite untenable. Instead it is my assumption that the high degree of order implied by Fibonacci parastichy counts can best be explained by a local mechanism repeatedly acting under the constraints of a slowly changing combination of genetic control and preexisting pattern, and that there can in principle be a decent model representation of this process which locally produces lattices.

11.2.3 Universality Across Phyla

Our most robust understanding of the molecular mechanism of Fibonacci phyllotaxis comes of course from *Arabidopsis*, while the most useful pattern observations come from the *Compositae*. Of course there are very many other phyllotactic patterns across the huge variety of land plants and beyond: there is evidence of Fibonacci structure in the bryophytes for example [49]. As Cronk wrote 'it would be extremely instructive to examine cases in which phyllotactic patterns change within a plant...and to correlate this with hormonally driven phase changes in plant development. It would be instructive too to consider how general the primordium auxin sink model is. The early microsurgical experiments [129] were done with ferns, particularly Dryopteris, and the results are consistent with this model. However the microphylls of lycophytes and the phyllidia of bryophytes have precisely determined phyllotaxy and it would be interesting to know how similar the determining system is' [29, p. 120]. While most plant molecular biologists are only recently leaving the constraints of *Arabidopsis*, modellers are under no such constraint. There is an opportunity to evaluate how adequate node-placement models are to explain observed pattern across a wide range of plant phyla.

11.3 Conclusion

I wrote this book because I wanted to understand what the simple reason was for Fibonacci structure in lattice-like patterns of nodes in plants. I have been surprised to discover that one of the biggest gaps in my understanding is not what the most appropriate biological model for node placement is, nor why lattices consistently transform through Fibonacci pairs, but a lack of understanding of how those models might deliver lattices at all. Rarely in mathematical biology, that is simultaneously a mathematical and a biological problem.

Kuhlemeier's recent review of phyllotaxis concluded that 'regular patterning is simply an emergent property of the molecular mechanism of lateral inhibition', and that this was 'unsatisfactory' [67]. The notion of an emergent property is a very slippery one, which theoretical biology has long struggled with: too useful to do without but far too often impossible to rigorously characterise. As a mathematical biologist I differ from Kuhlemeier about the degree of satisfaction mathematical biology can bring here. Fibonacci phyllotaxis is for once an emergent property in a mathematically rigorous, biologically consistent framework which generates novel hypotheses to advance scientific understanding.

The grand biological project of the twenty-first century is to move from knowledge of the genome, a list of named DNA sequences, to understanding of the organism, an interaction of functional systems. But genes are not mathematical objects: not only is there no gene that codes the number 55, still less can there be one that encodes the irrational golden ratio. It will not be possible to fully describe these common, robust, developmental pathways of morphogenesis without the use of at least some mathematical structure to understand 'emergent' pattern. Fibonacci phyllotaxis is an existence proof for the necessity of a mathematical systems biology, with suitable modesty and grounding in the culture of biology, to play a role in this grand project.

Chapter 12
Answers to Exercises

1.1 The grocery proof is left as an exercise to the reader. If using the Linford pineapple, there are clearly 8 spirals going up and to the right; depending on how the scales as the edges were originally joined there are either 12 or 13 spirals up and to the left.

1.2 Based on the consistent jumps between adjacent numbered scales (e.g. in Fig. 1, from 22 to 26 to 40 to 34 is evidence of four separate spirals), I would say these cones display evidence of 4 and 7 spirals (Fig. 1); 7 and 11 (Fig. 2); 7 and 10 (Fig. 4a). Based on the scale numbering alone, I can't assign an unambiguous spiral count for Fig. 3.

3.1 Some winding-number pairs are given in Table 12.1.

3.2 Multiply (3.9) by g.

3.3 It is easy to show by induction that

$$\begin{pmatrix} F_{j+1} & F_j \\ F_j & F_{j-1} \end{pmatrix} = (E^{-1}S)^j. \tag{12.1}$$

So the q_i's are all 1 and the determinant shows that (F_j, F_{j-1}) is a Bézout pair. It is not a winding-number pair because $F_{j-1}/F_j > \frac{1}{2}$. Note these are not the matrices that appear in Fig. 3.1.

3.4 Compute the difference between the Farey sum and the endpoints. the signs of these and the difference between the endpoints are all controlled by the sign of Δ.

3.5

$$ES \begin{pmatrix} F_{j+1} & F_{j-1} \\ F_j & F_{j-2} \end{pmatrix} = \begin{pmatrix} 1 & 1 \\ 1 & 0 \end{pmatrix} \begin{pmatrix} F_{j+1} & F_{j-1} \\ F_j & F_{j-2} \end{pmatrix} = \begin{pmatrix} F_{j+2} & F_{j+1} \\ F_{j+2} & F_{j-1} \end{pmatrix} \tag{12.2}$$

Table 12.1 Winding number pairs given as Farey intervals $[u/m, v/n]$. For m and n positive and distinct these are all contained in $[0, \frac{1}{2}]$ but the natural order of the endpoints varies with the sign of $mv - nu = \pm 1$

$\left[\dfrac{u}{m}, \dfrac{v}{n}\right]$	$\left[\dfrac{0}{1}, \dfrac{1}{0}\right]$			
$\left[\dfrac{1}{0}, \dfrac{0}{1}\right]$	$\left[\dfrac{1}{1}, \dfrac{0}{1}\right]$	$\left[\dfrac{1}{2}, \dfrac{0}{1}\right]$	$\left[\dfrac{1}{3}, \dfrac{0}{1}\right]$	$\left[\dfrac{1}{4}, \dfrac{0}{1}\right]$
$\left[\dfrac{0}{1}, \dfrac{1}{2}\right]$		$\left[\dfrac{1}{3}, \dfrac{1}{2}\right]$		
$\left[\dfrac{0}{1}, \dfrac{1}{3}\right]$	$\left[\dfrac{1}{2}, \dfrac{1}{3}\right]$		$\left[\dfrac{1}{4}, \dfrac{1}{3}\right]$	
$\left[\dfrac{0}{1}, \dfrac{1}{4}\right]$		$\left[\dfrac{1}{3}, \dfrac{1}{4}\right]$		
$\left[\dfrac{0}{1}, \dfrac{1}{5}\right]$	$\left[\dfrac{1}{2}, \dfrac{2}{5}\right]$	$\left[\dfrac{1}{3}, \dfrac{2}{5}\right]$	$\left[\dfrac{1}{4}, \dfrac{1}{5}\right]$	
$\left[\dfrac{0}{1}, \dfrac{1}{6}\right]$				
$\left[\dfrac{0}{1}, \dfrac{1}{7}\right]$	$\left[\dfrac{1}{2}, \dfrac{3}{7}\right]$	$\left[\dfrac{1}{3}, \dfrac{2}{7}\right]$	$\left[\dfrac{1}{4}, \dfrac{2}{7}\right]$	
$\left[\dfrac{0}{1}, \dfrac{1}{8}\right]$		$\left[\dfrac{1}{3}, \dfrac{3}{8}\right]$		
$\left[\dfrac{0}{1}, \dfrac{1}{9}\right]$	$\left[\dfrac{1}{2}, \dfrac{4}{9}\right]$		$\left[\dfrac{1}{4}, \dfrac{2}{9}\right]$	

3.6 For $k = 1$, the mediant of the Farey interval $(u + v)/(m + n)$ is F_j/F_{j+2}. Generalised Fibonacci number pairs, such as the Lucas numbers $F_4^3 = 7$ and $F_5^3 = 11$, also appear in Fig. 3.1. From inspection of the tree or by induction we have for $k > 1$

$$\begin{pmatrix} F_{j+1}^k & F_j \\ F_j^k & F_{j-1} \end{pmatrix} = (E \cdot S)^{j-1} \cdot (E^k S). \tag{12.3}$$

(This is still true for $k = 1$ but as we have seen delivers a Bézout pair but not the winding-number pair). For these generalised pairs the Farey mediant is

$$\frac{F_j + F_{j-1}}{F_{j+1}^k + F_j^k} = \frac{F_{j+1}}{F_{j+2}^k} \tag{12.4}$$

$$= \frac{F_{j+1}}{F_j + k F_{j+1}} \tag{12.5}$$

$$= \frac{F_{j+1}/F_j}{1 + k F_{j+1}/F_j} \tag{12.6}$$

12 Answers to Exercises

making use of the well-known relation $F_{j+1}^k = kF_j + F_{j-1}$. Finally we use $\lim_{k\to\infty} F_{j+1}/F_j = \tau$ together with the fact that the Farey intervals are nested.

4.1 If $\mathbf{p}_k = (x, z)$, then $|x| \leq \frac{1}{2}$ direct from the definition of a parastichy vector. If $x > 0$, then $\hat{\mathbf{p}}_k = \mathbf{p}_k - \mathbf{p}_0$ with horizontal component $x - 1$ which is between -1 and $-\frac{1}{2}$, and similarly for $x < 0$. Adding x components gives the final sentence.

4.2 For $d > \frac{1}{4}$, $\mathbf{p}_2 = (2d - 1, h) = 2\mathbf{p}_1 - (1, 0) \neq 2\mathbf{p}_1$; in general we need to set $w_k = [kd]$. For the second counter-example, choose d so that $\mathbf{p}_1$ and $\mathbf{p}_2$ both have a positive x component and $\mathbf{p}_3$ has a negative one.

4.3 Each time we move from ⓚ to ⓚ₊₁ we move to the next label in order of the foliation, which has m members. When we have moved m times we return to the original member.

4.4 If the pair (x_m, mh) and (x_n, nh) is generating in the lattice $\mathcal{L}(d, h)$, then the lattice $\mathcal{L}(d, sh)$ has a generating pair (x_m, smh) and (x_n, snh).

If $\mathbf{p}_m$ and $\mathbf{p}_n$ are a generating pair for $\mathcal{L}$ then so is any invertible linear combination

$$\begin{pmatrix} \mathbf{p}'_m \\ \mathbf{p}'_n \end{pmatrix} = \begin{pmatrix} a & b \\ c & d \end{pmatrix} \begin{pmatrix} \mathbf{p}_m \\ \mathbf{p}_n \end{pmatrix} \qquad (12.7)$$

for integers a, b, c, d. Most of these will not be parastichy vector pairs.

4.5 Any parastichy vector $\mathbf{p}_k$ can be written as $k\mathbf{p}_1 - w_k\mathbf{p}_0$, so $(0, 1)$ is generating. The rise of every integer sum of $\mathbf{p}_0$ and $\mathbf{p}_{n>1}$ is a multiple of nh and cannot be equal to h, and the pair cannot generate $\mathbf{p}_1$.

4.6 Since $\mathbf{p}_0 = \mathbf{p}_1 \pm \hat{\mathbf{p}}_1$, every parastichy vector $\mathbf{p}_k = k\mathbf{p}_1 - w\mathbf{p}_0$ is an integer sum of $\mathbf{p}_1$ and $\hat{\mathbf{p}}_1$. However every integer sum of $\mathbf{p}_m$ and $\hat{\mathbf{p}}_m$ has rise which is a multiple of m and so the pair is not generating unless $|m| = 1$.

4.7 The only candidate vector sum with the correct rise is $2\mathbf{p}_7 - 2\mathbf{p}_5$, but by direct calculation this is $\mathbf{p}_4 - (1, 0)$ and thus is not a parastichy vector.

4.8 $\mathbf{p}_m$ lies on the m-parastichy through the origin, and on some n-parastichy. If this was not the one adjacent to the n-parastichy through the origin, then it would intersect that n-parastichy and the intersection would be a lattice point since the pair is generating. But then the intersection would be a lattice point in but not at the corners of the m, n parallelogram which it cannot be since the pair is generating.

4.9 From Theorem 4.3, $h\Delta_{1n} = \mathbf{p}_1 \times \mathbf{p}_n = [nd]$. Setting $\sigma = \text{sign}(nd - [nd])$, $\hat{\mathbf{p}}_n = \mathbf{p}_n - \sigma\mathbf{p}_0$ and so $\Delta_{1\hat{n}} = [nd] + \sigma$. If $0 < d < \frac{1}{4}$ then $\Delta_{12} = 1$ and $\Delta_{1\hat{2}} = 0$, while if $\frac{1}{4} < d < \frac{1}{2}$ then $\Delta_{12} = 0$ and $\Delta_{1\hat{2}} = 1$. Finally we must choose u, v so that $|mv - u(n + \sigma)| = 1$ and

$$\begin{pmatrix} \hat{\mathbf{p}}_n \\ \mathbf{p}_m \end{pmatrix} = \begin{pmatrix} (n + \sigma) & -v \\ m & -u \end{pmatrix} \begin{pmatrix} \mathbf{p}_1 \\ \mathbf{p}_0 \end{pmatrix} \qquad (12.8)$$

4.10 By Exercise 4.9 the generating and complementary-generating intervals for $m = 0$ and $n = 1$ are the whole d-interval $[0, \frac{1}{2}]$. The generating interval for 1, 1 is empty, since $(1,1)$ is never generating, but the complementary-generating interval is the whole d-interval, since $(1,\hat{1})$ is always generating. For $m = 1, n = 2$, Exercise 4.9 shows that the generating interval is $[\frac{1}{4}, \frac{1}{2}]$ and the complementary-generating interval is the whole d-interval. For $m = 1$ and $n > 2$, Exercise 4.9 shows the generating interval is the d such that $[nd] = 1$, which is $2nd \in [1, 3]$. The complementary pair is generating either when (a) $[nd] = 0$, which requires $2nd < 1$ or (b) when both $[nd] = 2$ (so $2nd \in [3, 5]$) and $nd < [nd] = 2$, so (b) requires $2nd \in [3, 4]$. These two possibilities flank the generating interval.

4.11 Write the intervals for $\Delta = -1$ as $(L_{m-}, R_{m-}) = (u_- - \frac{1}{2}, u_- + \frac{1}{2})/m$, $(L_{n-}, R_{n-}) = (v_- - \frac{1}{2}, v_- + \frac{1}{2})/n$. Use $u_- = m - u$ and $v_- = n - v$.

4.12 Suppose k is odd, then the winding-number pair for $m = F_k, n = F_{k+1}$ is $u = F_{k-1}, v = F_k$. Let I_k be the interval $[F_k - \frac{1}{2}, F_k + \frac{1}{2}]/F_{k+1}$. Then $[md] = u$ when $d \in I_k$ and $[nd] = v$ when $d \in I_{k+1}$. As k becomes large and odd, I_k tends to the point $\lim F_k/F_{k+1} = 1/\tau$. If k is even, the winding-number pair is $F_{k-2}, F_{k-1} = F_k - F_{k-1}, F_{k+1} - F_k$ which we recognise as giving us the interval $1 - I_k$ for $[md] = u$ and $[nd] = v$ instead. Since $1/\tau > \frac{1}{2}$, in either case the interval in $[0, \frac{1}{2}]$ on which parastichy vectors for adjacent Fibonacci numbers is generating is close to $1 - 1/\tau = 1/\tau^2$.

4.13 In Theorem 4.5 the vectors are restricted to be parastichy vectors, but to find two generating and opposed vectors with $m = 1$ and $n = 1$ we must allow complementary vectors and take $\mathbf{p}_1$ and $\hat{\mathbf{p}}_1$. These are always generating and opposed and so the generating interval is $[0, 1]$. Either Bézout pair of $u = 1, v = 0$ or $u = 0, v = 1$ allows this to be written as $[u/m, v/n]$ in some order.

Suppose that $m = 1$ and $n > 1$. If $dn < \frac{1}{2}$, then $\mathbf{p}_1$ and $\mathbf{p}_n$ are co-linear, so instead pay attention to the pair $\mathbf{p}_1$ and $\hat{\mathbf{p}}_n$. For d small and positive this pair is generating and opposed, and remains so until $dn = \frac{1}{2}$ at which point the horizontal component of $\mathbf{p}_n$ increases through $\frac{1}{2}$ and jumps back down to $-\frac{1}{2}$ so that $\mathbf{p}_n$ is no longer co-linear with $\mathbf{p}_1$. As d continues to increase, the new pair $\mathbf{p}_1$ and $\mathbf{p}_n$ remain generating and opposed until $\mathbf{p}_n$ becomes vertical, which happens when $nd = 1$. After this the pair is no longer opposed. For $m = 1$ amd $n > 1$, the winding-number pair is $u = 0$ and $v = 1$, corresponding to a Farey interval of $[0, 1/n]$. Thus this is the interval on which the lattice has an opposed generating pair, but below the midpoint $d < 1/2n$, the pair is of a parastichy vector and a complementary vector.

4.14 We set $m = 55 = F_j$ and $89 = n = F_j$ with $j = 10$, and use $u = F_9 = 34$, $v = F_{10} = 55$ to give $mv - nu = (-1)^{11}$ as expected. The interval in $[0, \frac{1}{2}]$ is $1 - (34/55, 55/89) = 1/\tau^2 + [-0.00014\ldots, 0.0005\ldots]$. Thus the divergence angle of such a lattice must be within less than 2 parts in a thousand of the golden angle.

4.15 For h large enough, the shortest vector in the lattice is $\mathbf{p}_0 = (1, 0)$ of length 1, and the second shortest is $\mathbf{p}_1 = (h, d)$ of length $h^2 + d^2$. These change relative

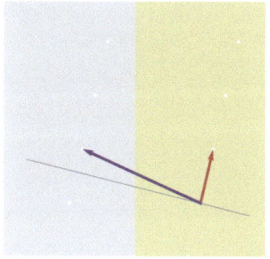 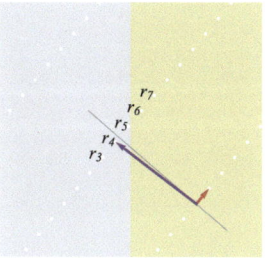 

Fig. 12.1 First (red) and second (blue) parastichies in a lattice with $d = 7/72$ and (left to right) h equal to $0.4, 0.115, 0.08$. The corresponding principal parastichy pairs are $(1, \hat{1})$, $(1, \hat{3})$, and $(1, 6)$. The normal to the 1-parastichy is shown as a thin line. For large enough h, as in the first two cases, one of the principal vectors is a complementary vector.

magnitude when $h^2 + d^2 = 1$ giving us a $(1, 0)$ lattice. As h decreases further, there is a point when $\hat{\mathbf{p}}_1 = (h, d - 1)$ is of equal length to $\mathbf{p}_0$: this is when $h^2 + d^2 = 2d$. These give the boundaries between the $(0, 1)$, $(1, 0)$ and $(1, \hat{1})$ regions in Fig. 5.1.

As h continues to decrease, but as long as $\mathbf{p}_1$ remains the shortest vector in the lattice, the second shortest must be on the adjacent 1-parastichy to the origin. Given the definition of $k = \lfloor 1/2d \rfloor - 1 \geq 1$, $1/2(k + 1) < d \leq 1/2k$, and then $\mathbf{p}_2, \ldots, \mathbf{p}_k$ are co-linear with $\mathbf{p}_1$ on the origin parastichy, so they are not the second shortest principal vector. On the adjacent 1-parastichy, though, we can set $\mathbf{r}_n = (nd - 1, nh)$, and we have $\hat{\mathbf{p}}_n = \mathbf{r}_n = \mathbf{p}_n - \mathbf{p}_0$ for $-k \leq n \leq k$ and $\mathbf{p}_n = \mathbf{r}_n$ for $k + 1 \leq n \leq 2k + 1$. Then the second principal vector is one of these $\mathbf{r}$s and specifically it is the one closest to $\mathbf{n}$: the vector which normal to the origin 1-parastichy from the origin to the adjacent 1-parastichy: see Fig. 12.1.

We have $\mathbf{p}_1 = (d, h)$; so $\mathbf{n}$ has slope $-d/h$. Since $\mathbf{n}$ and $\mathbf{p}_1$ must form a rectangle of area h, we can find the length of $\mathbf{n}$ and discover $\mathbf{n} = \nu(-h^2/d, h)$ where $\nu = d/(h^2 + d^2)$ Since the rise of $\mathbf{n}$ is $h\nu$, it passes through one of the $\mathbf{r}_n$ every time ν passes through an integer, and the closest $\mathbf{r}_n$ has $n = [\nu]$, so that the second principal vector changes from $\mathbf{r}_n$ to $\mathbf{r}_{n+1}$ when $\nu = n + \frac{1}{2}$ which can be rewritten as

$$h^2 + \left(d - \frac{1}{2n+1}\right)^2 = \frac{1}{(2n+1)^2}. \tag{12.9}$$

Note that for the initial case $n = 0$ we recover $h^2 + d^2 = 2d$ as the $0 = \hat{1}$ boundary where a $(1, 0)$ lattice became a $(1, \hat{1})$ lattice. For each of $n = 1, \ldots, k$ this gives the point at which a $(1, \hat{n})$ becomes a $(1, \widehat{n+1})$ lattice, and then for $n = k + 1$ the lattice transitions from a $(1, \hat{k})$ to a $(1, k + 1)$ lattice. From this point on the principal pair are both parastichy vectors, though at some yet smaller h, $\mathbf{p}_1$ will cease to be the principal vector and we are back in the full complexity of Fig. 5.1.

The region of lattice space in which lattices are of the form $(1, \hat{n})$ is shown in Fig. 12.2.

Fig. 12.2 Structure of lattices space near $(d, h) = (0, 0)$ where lattices are spiral lattices. Shaded in light and dark yellow are regions where the principal parastichy pair are $(1, \hat{n})$ for $n = 1$ to $n = 7$. At each of the light-dark boundaries the lattice is square, as will be shown in the next chapter. The principal pair transitions from $(1, \hat{n})$ to $(1, n)$ at the vertical line $d = \frac{1}{2n}$

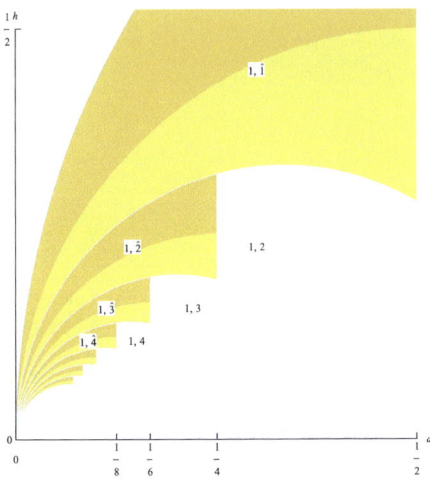

4.16 The pair (73,103) is successively reduced to (30,73), (13,30), and terminates at (4,13) which is the principal pair for the lattice.

4.17 If the angle between any two parastichy vectors was not 60°, then the third of the vectors, which must be their sum or difference, would not be the third side of an equilateral triangle and would not also have the same length.

5.1 Note that $nx_n - mx_n = \Delta$ and use this to eliminate x_n and r^2 from (5.1) and (5.2). This (or alternatively eliminating x_m) yields

$$r^2 = \frac{2\Delta n x_m - 1}{4(n^2 - m^2)} \tag{12.10}$$

$$r^2 = \frac{2\Delta m x_n + 1}{4(n^2 - m^2)}. \tag{12.11}$$

5.2 Substituting for x_m and x_n into (5.1) and (5.2) and eliminating r^2 gives

$$d^2 m^2 + 2dnv + h^2 m^2 + u^2 = d^2 n^2 + 2dmu + h^2 n^2 + v^2. \tag{12.12}$$

Using the definition of $\bar{d}$ and $\Delta^2 = 1$ this gives (5.5). Equation (5.6) follows from the answer to Exercise 5.1, or by using $\Delta = mv - nu$ to eliminate v in (12.12) and then eliminating h^2 using (5.1).

5.3 Suppose $m < n$. The d, h semi-circle intersects the $h = 0$ axis at $(d_L = (v - u)/(n - m), 0)$, with $x_m = x_n$, and $(d_R = (v + u)/(n + m), 0)$, and $(n^2 - m^2)(d_L - d_R) = 2\Delta$. At these points $2r(n \pm m) = 1$. Except in the cases $(0 = 1)$, $(1 = 1)$ and $(1 = 2)$, d_L and d_R either both lie in $[0, \frac{1}{2}]$ or both lie in $[\frac{1}{2}, 0]$.

5.4 Consider the triangle of Fig. 4.8 which has sides $n\mathbf{p}_m$, $m\mathbf{p}_n$ and 1. Since the lattice is hexagonal, the top angle of the triangle there is 60°, so set $|\mathbf{p}_m| = |\mathbf{p}_n| = 2r$ and use the cosine rule to find

$$4r^2 = \frac{1}{m^2 + mn + n^2}.$$

With this form, equations (12.10) and (12.11) simplify to

$$x_m = 2\Delta(m + 2n)r^2$$
$$x_n = -2\Delta(2m + n)r^2$$
$$x_{m+n} = 2\Delta(n - m)r^2$$

and by using $(n^2 - m^2)h^2 = x_m^2 - x_n^2 = 3 \cdot 4r^4(n^2 - m^2)$ we get equation (5.7) and could further use this to find $d - \bar{d}$ as a function of m and n. Note this shows that $\mathbf{p}_m$ and $\mathbf{p}_n$ are opposed and that none of the principal vectors are vertical at a triple point with $n > m$.

5.5 A square lattice occurs for $\mathbf{p}_m \cdot \mathbf{p}_n = 0$; using $|n\mathbf{p}_m - m\mathbf{p}_n| = 1$ we get $1 = n^2|\mathbf{p}_m|^2 + m^2|\mathbf{p}_n|^2 = 4r^2(m^2 + n^2)$. This is unique on the branch because $4r^2$ is strictly decreasing.

5.6 At the triple point $m = n = m + n$, the m=n branch, interpreted as a function $h(d)$, has a slope h' given by

$$\frac{h'}{h} = -\frac{1}{6r^2\Delta}. \tag{12.13}$$

5.7 For the original lattice in (0,1), we have $h_0 \geq 0$. Separation into real and imaginary parts gives

$$z_1 = g_{mn}(d_0 + ih_0) \tag{12.14}$$

$$\Re z_1 = \frac{nv + mu|w^2| - d_0(nu + mv)}{|mw - n|^2} \tag{12.15}$$

$$\Im z_1 = \frac{h_0 \Delta_{mn}}{|mw - n|^2} \tag{12.16}$$

and then setting $h = \Im \Delta_{mn} z_1$ shows that z has positive rise.

5.8 Parameterise the (0,1) region of van Iterson space by setting $w = d_0 + ih_0$ and choose a transformed lattice with $z = \Delta g_{mn}(w)$ in (12.16) gives

$$\Re z = \Delta \frac{nv + mu|w^2| - d_0(nu + mv)}{|mw - n|^2} \tag{12.17}$$

$$\Im z = h_0 \frac{1}{|mw - n|^2} = \frac{h_0}{n^2 + m^2|w|^2 - 2mnd_0}. \tag{12.18}$$

The orthogonal lattices in $(0,1)$ are $\mathcal{L}(is)$ for $0 < s < 1$, so we set $d_0 = 0$, $h_0 = s$ and $|w|^2 = s^2$ to get

$$d(s) = \frac{mu + nvs^2}{n^2 + m^2 s^2}, \tag{12.19}$$

$$h(s) = \frac{s}{n^2 + m^2 s^2}. \tag{12.20}$$

They must lie on a circle because the Möbius map takes vertical lines in the complex plane into circles and indeed

$$\left(d(s) - \frac{1 + 2nu}{2mn}\right)^2 + h(s)^2 = \frac{1}{4m^2n^2}. \tag{12.21}$$

5.9 Touching circle lattices occur in $(0,1)$ space when $w = e^{i\theta}$, for $\pi/3 \le \theta \le 2\pi/3$ and θ is the angle between the principal vectors. Equations (12.17) and (12.18) give the divergence and rise of the lattice transformed by Δg_{mn} as

$$d(\theta) = \Delta \frac{nv + mu - \cos\theta(nu + mv)}{m^2 - 2\cos\theta mn + n^2} \tag{12.22}$$

$$h(\theta) = \frac{\sin\theta}{m^2 - 2\cos\theta mn + n^2}. \tag{12.23}$$

For a square touching-circle lattice, $\theta = \pi/2$ and

$$d = \Delta \frac{mu + nv}{m^2 + n^2}, \tag{12.24}$$

$$h = \frac{1}{m^2 + n^2}. \tag{12.25}$$

There are two hexagonal lattice points in $(0,1)$ at $e^{i\pi/3}$ and $e^{2i\pi/3}$ which map to different hexagonal lattices corresponding to principal parastichies $(m, n, n + m)$ or $(m, n - m, n)$. For $w = e^{i\pi/3}$ we get

$$d = \Delta \frac{mu + nv - (nu + mv)/2}{m^2 + mn + n^2}, \tag{12.26}$$

$$h = \frac{\sqrt{3}}{2} \frac{1}{(m^2 - mn + n^2)}. \tag{12.27}$$

5.10 The angle θ between the principal parastichy vectors is the same in the $(0,1)$ lattice and in the rotated and scaled lattice, and $w = |w_0|e^{i\theta}$. Inserting this into in (12.16) and taking real and imaginary parts allows the calculation of $\mathbf{p}_m = mz_1 - u$ and $|\mathbf{p}_m|^2$:

$$|\mathbf{p}_m|^2 = \frac{1}{m^2 w_0^2 - 2mn w_0 \cos(\theta) + n^2}. \tag{12.28}$$

For a touching-circle lattice, $w_0 = 1$ and $4r^2 = |\mathbf{p}_m|^2$ so the radius of the touching circles (not the radius of the circle in van Iterson space where lattices are touching circles!) is

$$4r^2 = \frac{1}{m^2 - 2mn\cos(\theta) + n^2}. \tag{12.29}$$

$$= \frac{h(\theta)}{\sin\theta} \tag{12.30}$$

Eliminating $\cos\theta$ between this and (12.23) gives

$$d = \tfrac{1}{2}\left(\tfrac{u}{m} + \tfrac{v}{n}\right) + 2\Delta\left(\tfrac{n}{m} - \tfrac{m}{n}\right)r^2. \tag{12.31}$$

As w_1 rotates around the unit circle from $e^{2i\pi/3}$ to $e^{i\pi/2}$ to $e^{i\pi/3}$, the inverse square diameter of the transformed lattice increases from $m^2 - mn + n^2$ at the top triple point on the $m = n$ branch, through $m^2 + n^2$ at the square lattice point, down to $m^2 + mn + n^2$ at the lower triple point.

5.11 If m and n are adjacent Fibonacci numbers F_k and F_{k+1}, then using $F_k \approx \tau^k/\sqrt{5}$ we find the value of r on the branch is approximately

$$F_k\sqrt{2} \leq 2r^{-1} \leq F_k\sqrt{2(1+\tau)}. \tag{12.32}$$

So on the Fibonacci branch of the van Iterson diagram, the larger Fibonacci number becomes F_{k+1} at approximately $r = \sqrt{5/2}\tau^{-k} \approx \frac{1}{\sqrt{2}F_k}$.

5.12 From (12.23), the imaginary part of z_1 is zero when $\sin\theta = 0$, so the left and right intersections of the image of w on the real axis are in some order $(u-v)/(m-n)$ and $(u+v)/(m+n)$ so that the radius of the van Iterson circle is

$$v_r = \frac{1}{n^2 - m^2} \tag{12.33}$$

and it is centred at

$$v_0 = \frac{|mu - nv|}{n^2 - m^2} \tag{12.34}$$

so that $z_1 = \Delta v_0 + v_r e^{i\psi}$. A bit more work shows

$$\tan \psi = \frac{(n^2 - m^2)\sin\theta}{(m^2 + n^2)\cos\theta - 2mn} \tag{12.35}$$

$$= \frac{(n^2 - m^2)}{n^2 - 2mn/\cos\theta + m^2} \tan\theta. \tag{12.36}$$

The hexagonal points are when $\cos\theta = \pm\sqrt{3}/2$ and $\sin\theta = 1/2$, so

$$\tan\psi = \frac{(n^2 - m^2)}{\pm\sqrt{3}(n^2 + m^2) - mn}. \tag{12.37}$$

5.13 The bifurcation theory argument of this chapter or specifically examination of Fig. 5.5 shows we must have $n - m > m$ or $m < \frac{1}{2}n$. If this is true, the lattice is non-opposed on the (m=n) branch from the triple point (m=n=n-m) to the point where either $\mathbf{p}_m$ or $\mathbf{p}_n$ vector becomes vertical.

Choose u, v as the winding-number pair so that $mv - nu = \Delta$. We need either $\Re\mathbf{p}_m = 0$, so $\Re z_1 = u/m$ or $\Re\mathbf{p}_n = 0$, so $\Re z_1 = v/n$. This gives $\cos\theta = n/m$ or $\cos\theta = m/n$ respectively, so we need to choose $\cos\theta = m/n$, and since $m/n < \frac{1}{2}$, $\pi/3 \le \theta \le 2\pi/3$ and one of the solutions of $\cos\theta = m/n$ is indeed a point on the touching-circle branch. From the previous exercise $\tan\psi = -\tan\theta$ at that point, and so $|\cos\psi| = m/n$ there too.

5.14 Lagrange proved in 1773 that a hexagonal lattice is the closest possible lattice packing, and triple points are the only points in lattice space at which the lattice is hexagonal. For more on the connection between the modular group and sphere-packings, see, say, Berger [11].

6.1 The first result itself follows from reading the rise of a lattice from (12.16), combined with observing that Fibonacci numbers scale like τ^k and so h scales like n^{-2}.

The exercise is to warn the reader against trying to make too much sense of the following strange expression which appeared in [96]:

$$k \approx \text{P.I} = 0.33 - 2.39 \log_{10} \log_{10} R. \tag{12.38}$$

Richards wanted to fit real disk patterns as idealisations of those on the left of Fig. 6.3. He seems to have assumed an orthogonal golden lattice transformed with logarithmic spirals, and his unfortunate idea was to use the R of equation (6.4) to estimate h, which gives one log in Eq. 12.38, and then use the knowledge of how h scaled with k to estimate k, giving another log in that equation. The specific numerical values of the coefficients in equation (6.4) depend on exactly what lattice is being imagined as the precursor. The argument in [96] is not easy to follow mathematically, creating an opportunity for several subsequent papers to attempt a more rigorous formulation [58, 118, 120].

Quite apart from the problems inherent in a log log, term, estimates of R will be in practice very sensitive to the estimate of the position of the unknown centre of the disk. Even if the plastochrone ratio R can be reliably estimated from the observed pattern, then it seems much, much, simpler to just count the k. But more fundamentally the Richards plastochrone index just assumes too much to discriminate between different developmental models.

References

1. I. Adler. 'A Model of Contact Pressure in Phyllotaxis'. *Journal of Theoretical Biology* 45.1 (1974), pp. 1–79. https://doi.org/10.1016/0022-5193(74)90043-5.
2. I. Adler. 'The Consequences of Contact Pressure in Phyllotaxis'. *Journal of Theoretical Biology* 65.1 (1977), pp. 29–77. https://doi.org/10.1016/0022-5193(77)90077-7.
3. E. Aliyev and K. Vedmedieva. 'Mathematical Model of Placement of Sunflower Seeds in a Head'. *Scientific and Technical Bulletin of the Institute of Oilseed Crops NAAS* 34 (2023), pp. 15–23. URL: https://bulletin.imk.zp.ua/index.php?menu=4&id=449&lang=en.
4. E. Aliyev et al. 'Study of the Distribution of Phenotypic Characteristics of Sunflower Seeds in a Head of Different Genotypes'. *Bulgarian Journal of Crop Science* (29th Aug. 2024). https://doi.org/10.61308/JIXX8922. URL: https://cropscience-bg.org/page/en/details.php?article_id=1193 (visited on 03/09/2024).
5. E. Artin. 'Ein mechanisches System mit quasiergodischen Bahnen'. *Abhandlungen aus dem Mathematischen Seminar der Universität Hamburg* 3.1 (1st Dec. 1924), pp. 170–175. ISSN: 1865-8784. https://doi.org/10.1007/BF02954622. URL: https://doi.org/10.1007/BF02954622 (visited on 28/01/2025).
6. P. Atela. 'The Geometric and Dynamic Essence of Phyllotaxis'. *Mathematical Modelling of Natural Phenomena* 6.02 (2011), pp. 173–186. https://doi.org/10.1051/mmnp/20116207.
7. P. Atela and C. Golé. 'Rhombic Tilings and Primordia Fronts of Phyllotaxis'. 2017. https://doi.org/10.48550/arXiv.1701.01361. arXiv: 1701.01361.
8. P. Atela, C. Golé and S. Hotton. 'A Dynamical System for Plant Pattern Formation: A Rigorous Analysis'. *Journal of Nonlinear Science* 12.6 (2002), pp. 641–676. https://doi.org/10.1007/s00332-002-0513-1.
9. C. J.-C. Ballot and H. C. Williams. *The Lucas Sequences: Theory and Applications*. Springer Nature, 2023. 312 pp. ISBN: 978-3-031-37238-4.
10. D. Barabé and C. Lacroix. *Phyllotactic Patterns: A Multidisciplinary Approach*. World Scientific, May 2020. ISBN: 978-981-12-1100-3. URL: https://www.worldscientific.com/worldscibooks/ https://doi.org/10.1142/11571.
11. M. Berger. *Geometry Revealed: A Jacob's Ladder to Modern Higher Geometry*. Springer, 2010. ISBN: 978-3-540-70996-1.
12. K. van Berkel et al. 'Polar Auxin Transport: Models and Mechanisms'. *Development* 140.11 (2013), pp. 2253–2268. ISSN: 0950-1991, 1477-9129. https://doi.org/10.1242/dev.079111. pmid: 23674599. URL: https://dev.biologists.org/content/140/11/2253 (visited on 26/09/2019).
13. F. Besnard et al. 'Cytokinin Signalling Inhibitory Fields Provide Robustness to Phyllotaxis'. *Nature* (2014). https://doi.org/10.1038/nature12791. URL: https://www.nature.com/articles/nature12791 (visited on 05/03/2019).

14. C. Bonnet. *Recherches Sur l'usage Des Feuilles Dans Les Plantes*. Gottingen and Leiden: Luzac, 1754. https://doi.org/10.12345/6789. URL: https://www.erara.ch/zut/2303422.
15. A. Braun. *Betrachtungen über die Erscheinung der Verjüngung in der Naturinsbesondere in der Lebens- und Bildungsgeschichte der Pflanze*. Wilhelm Engelmann, 1851. 390 pp. URL: https://archive.org/details/betrachtungenbe00braugoog/page/n6/mode/2up.
16. A. Braun. 'Dr. Carl Schimper's Vorträge Über Die Moglichkeit Eines Wissenschaftlichen Verständnisses Der Blattstellung'. *Flora, oder Allgemeine Botanische Zeitung (Regensburg)* 18 (1835), 145-192 and Tab 1.
17. A. Braun. 'Vergleichende Untersuchung über die Ordnung der Schuppen an den Tannenzapfen als Einleitung zur Untersuchung der Blattstellungen Überhaupt'. *Nova acta physico-medica Academiae Caesareae Leopoldino-Carolinae Naturae Curiosum* 15 (1831), 195-402. URL: https://www.biodiversitylibrary.org/item/113880.
18. L. Bravais and A. Bravais. 'Essai Sur La Disposition Des Feuilles Curvisériées'. *Annales des Sciences Naturelles Botanique Seconde Série* 7 (1837), 42-110 and plates. URL: https://www.biodiversitylibrary.org/bibliography/72750.
19. K. Bull-Hereñu et al. 'Mechanical Forces in Floral Development'. *Plants* 11.5 (5 Jan. 2022), p. 661. ISSN: 2223-7747. https://doi.org/10.3390/plants11050661. URL: https://www.mdpi.com/2223-7747/11/5/661 (visited on 04/02/2025).
20. J. M. Burke et al. 'Genetic Analysis of Sunflower Domestication'. *Genetics* 161.3 (2002), pp. 1257–1267. ISSN: 1943-2631. https://doi.org/10.1093/genetics/161.3.1257. URL: https://doi.org/10.1093/genetics/161.3 1257 (visited on 08/06/2021).
21. L. A. Bursill, P. Ju Lin and F. XuDong. 'Spiral Lattice Concepts'. *Modern Physics Letters B* 1.5-6 (1987), pp. 195–206.
22. F. Cartenì et al. 'Modelling the Development and Arrangement of the Primary Vascular Structure in Plants'. *Annals of Botany* 114.4 (2014), pp. 619–627. ISSN: 0305-7364. https://doi.org/10.1093/aob/mcu074.pmid:24799440. URL: https://www.ncbi.nlm.nih.gov/pmc/articles/PMC4156123/ (visited on 11/02/2019).
23. D. H. Chitwood et al. 'Leaf Asymmetry as a Developmental Constraint Imposed by Auxin-Dependent Phyllotactic Patterning'. *The Plant Cell* 24.6 (1st June 2012), pp. 2318–2327. ISSN: 1040-4651. https://doi.org/10.1105/tpc.112.098798. URL: https://doi.org/10.1105/tpc.112.098798 (visited on 13/03/2025).
24. A. H. Church. *On the Relation of Phyllotaxis to Mechanical Laws*. London: Williams and Norgate, 1904. https://doi.org/10.5962/bhl.title.57125. URL: https://archive.org/details/cu31924000658470/page/n331/mode/2up.
25. M. Codaccioni. 'Étude Phyllotaxique d'un Lot de 200 Plants d'. Helianthus Annuus L. Cultivés En Serre'. *Comptes rendus Académie des sciences* 241(1955), pp. 1159–1159.
26. K. Conrad. 'Ideal Classes and SL2' (2024). URL: https://kconrad.math.uconn.edu/blurbs/gradnumthy/SL2classno.pdf.
27. T. J. Cooke. 'Do Fibonacci Numbers Reveal the Involvement of Geometrical Imperatives or Biological Interactions in Phyllotaxis?' *Botanical Journal of the Linnean Society* 150.1 (2006), pp. 3–24. https://doi.org/10.1111/j.1095-8339.2006.00490.x.
28. Y. Couder. 'Initial Transitions, Order and Disorder in Phyllotactic Patterns: The Ontogeny of *Helianthus Annuus*: A Case Study'. *Acta Societatis Botanicorum Poloniae* 67.2 (1998), pp. 129–150. https://doi.org/10.5586/asbp.1998.016.
29. Q. Cronk. *The Molecular Organography of Plants*. Oxford University Press, 2009. ISBN: 0-19-955035-2.
30. Y. Deb et al. 'Phyllotaxis Involves Auxin Drainage through Leaf Primordia'. *Development* 142.11 (2015), pp. 1992–2001. https://doi.org/10.1242/dev.121244.
31. A. M. Décaillot. 'Edouard Lucas (1842-1891): le parcours original d'un scientifique français dans la deuxième moitié du XIXè siècle'. Thèse de doctorat. Université Paris Descartes, 1999. 207; 146.
32. S. Douady. 'The Selection of Phyllotactic Patterns'. In: *Symmetry in Plants*. Ed. by R. V. Jean and D. Barabé. World Scientific, 1998, pp. 335–358. ISBN: 978-981-02-2621-3.

33. S. Douady and Y. Couder. 'Phyllotaxis as a Dynamical Self Organizing Process (Parts I, II, III)'. *Journal of Theoretical Biology* 178 (1996).
34. S. Douady and Y. Couder. 'Phyllotaxis as a Physical Self-Organized Growth Process'. *Physical Review Letters* 68.13 (1992), pp. 2098–2101. https://doi.org/10.1103/PhysRevLett.68.2098.
35. S. Douady and C. Golé. 'Fibonacci or Quasi-Symmetric Phyllotaxis. Part II: Botanical Observations'. *Acta Societatis Botanicorum Poloniae* 85.4 (31st Dec. 2016). ISSN: 2083-9480. https://doi.org/10.5586/asbp.3534. URL: https://pbsociety.org.pl/journals/index.php/asbp/article/view/asbp.3534 (visited on 15/02/2019).
36. S. Douady et al. *Do Plants Know Math?* Princeton University Press, 2024.
37. P. Elomaa and T. Zhang. 'Understanding Capitulum Development: Gerbera Hybrida Inflorescence Meristem as an Experimental System'. *Capitulum* 1.2 (29th Jan. 2022), pp. 53–59. ISSN: 2789-2786. https://doi.org/10.53875/capitulum.01.2.04.
38. R. O. Erickson. 'The Geometry of Phyllotaxis'. In: *The Growth and Functioning of Leaves*. Ed. by J. E. Dale and F. L. Milthorpe. Cambridge University Press, 1983. ISBN: 0-521-23761-0.
39. V. Fierz. 'Aberrant Phyllotactic Patterns in Cones of Some Conifers: A Quantitative Study'. *Acta Societatis Botanicorum Poloniae* 84.2 (3rd July 2015), pp. 261–265. ISSN: 2083-9480. https://doi.org/10.5586/asbp.2015.025. URL: https://pbsociety.org.pl/journals/index.php/asbp/article/view/asbp.2015.025 (visited on 05/03/2019).
40. L. R. Ford. *Automorphic Functions*. 2nd edition. New York: Chelsea, 1951. ISBN: 978-0-8218-3741-2.
41. D. H. Fowler. *The Mathematics of Plato's Academy: A New Reconstruction*. 2nd edition. Clarendon Press, 1999. ISBN: 978-0-19-850258-6.
42. M. Fuchs and J. U. Lohmann. 'Aiming for the Top: Non-Cell Autonomous Control of Shoot Stem Cells in Arabidopsis'. *Journal of Plant Research* 133.3 (1st May 2020), pp. 297–309. ISSN: 1618-0860. https://doi.org/10.1007/s10265-020-01174-3. URL: https://doi.org/10.1007/s10265-020-01174-3 (visited on 11/09/2021).
43. C. S. Galvan-Ampudia et al. 'Phyllotaxis: From Patterns of Organogenesis at the Meristem to Shoot Architecture'. *Wiley Interdisciplinary Reviews: Developmental Biology* 5 (2016), pp. 460–473. https://doi.org/10.1002/wdev.231. URL: https://hal.archives-ouvertes.fr/hal-01413095 (visited on 01/09/2019).
44. C. S. Galvan-Ampudia et al. 'Temporal Integration of Auxin Information for the Regulation of Patterning'. *eLife* 9 (7th May 2020), e55832. ISSN: 2050-084X. https://doi.org/10.7554/eLife.55832.
45. C. F. Gauss. 'Bemerkung zu den fragmenten uber Elliptische ModulFunctionen'. In: *Werke, Volume 8*. Cambridge University Press, 2011, pp. 102–105. ISBN: 978-1-139-05829-2. https://doi.org/10.1017/CBO9781139058292.
46. C. Godin, C. Golé and S. Douady. Phyllotaxis as Geometric Canalization during Plant Development . Development 147.19 (12th Oct. 2020), dev165878. ISSN: 1477-9129. https://doi.org/10.1242/dev.165878.
47. C. Golé and S. Douady. 'Convergence in a Disk Stacking Model on the Cylinder'. *Physica D: Nonlinear Phenomena* 403 (1st Feb. 2020), p. 132278. ISSN: 0167-2789. https://doi.org/10.1016/j.physd.2019.132278. URL: http://www.sciencedirect.com/science/article/pii/S0167278919305093 (visited on 19/01/2021).
48. C. Golé, J. Dumais and S. Douady. 'Fibonacci or Quasi-Symmetric Phyllotaxis. Part I: Why?' *Acta Societatis Botanicorum Poloniae* 85.4 (31st Dec. 2016). ISSN: 2083-9480. https://doi.org/10.5586/asbp.3533. URL: https://pbsociety.org.pl/journals/index.php/asbp/article/view/asbp.3533 (visited on 15/02/2019).
49. C. Gómez-Campo. 'Phyllotactic Patterns in Bryophyllum Tubiflorum,' *Harvard Botanical Gazette* 135 (1974), pp. 49–58.
50. P. B. Green and D. R. Baxter. 'Phyllotactic Patterns: Characterization by Geometrical Activity at the Formative Region'. *Journal of Theoretical Biology* 128.3 (1987), pp. 387–395. https://doi.org/10.1016/S0022-5193(87)80080-2.
51. P. B. Green, C. S. Steele and S. C. Rennich. 'Phyllotactic Patterns: A Biophysical Mechanism for Their Origin'. *Annals of Botany* 77.5 (1996), pp. 515–528. https://doi.org/10.1006/anbo.1996.0062.

52. Y. Guédon et al. 'Pattern Identification and Characterization Reveal Permutationsof Organs as a Key Genetically Controlled Property of Post-Meristematic Phyllotaxis'. *Journal of Theoretical Biology* 338 (7th Dec. 2013), pp. 94–110. ISSN: 1095-8541. https://doi.org/10.1016/j.jtbi.2013.07.026.
53. C. J. Harrison and J. L. Morris. 'The Origin and Early Evolution of Vascular Plant Shoots and Leaves'. *Philosophical Transactions of the Royal Society B: Biological Sciences* 373.1739 (5th Feb. 2018), p. 20160496. https://doi.org/10.1098/rstb.2016.0496. URL: https://royalsocietypublishing.org/doi/10.1098/rstb.2016.0496 (visited on 07/10/2021).
54. M. G. Heisler et al. 'Alignment between PIN1 Polarity and Microtubule Orientation in the Shoot Apical Meristem Reveals a Tight Coupling between Morphogenesis and Auxin Transport'. *PLOS Biology* 8.10 (2010), e1000516. ISSN: 1545-7885. https://doi.org/10.1371/journal.pbio.1000516. URL: https://journals.plos.org/plosbiology/article?id=10.1371/journal.pbio.1000516 (visited on 26/09/2019).
55. L. F. Hernández and J. H. Palmer. 'Regeneration of the Sunflower Capitulum after Cylindrical Wounding of the Receptacle'. *American Journal of Botany* 75.9 (1988), pp. 1253–1261. https://doi.org/10.2307/2444447.
56. S. Hotton et al. 'The Possible and the Actual in Phyllotaxis: Bridging the Gap between Empirical Observations and Iterative Models'. *Journal of Plant Growth Regulation* 25.4 (2006), pp. 313–323. https://doi.org/10.1007/s00344-006-0067-9.
57. M. Ingrouille and B. Eddie. *Plants: Diversity and Evolution*. Cambridge University Press, 2006. ISBN: 0-521-79097-2. URL: https://www.amazon.co.uk/Plants-Diversity-Evolution-Martin-Ingrouille/dp/0521790972 (visited on 17/02/2021).
58. R. V. Jean. 'A Rigorous Treatment of Richard's Mathematical Theory of Phyllotaxis.' *Mathematical Biosciences* 44 (1979), pp. 221–40.
59. R. V. Jean. 'Number-Theoretic Properties of Two-Dimensional Lattices'. *Journal of Number Theory* 29.2 (1988), pp. 206–223. https://doi.org/10.1016/0022-314X(88)90100-X.
60. R. V. Jean. *Phyllotaxis: A Systemic Study in Plant Morphogenesis*. Cambridge University Press, 1994. ISBN: 0-521-40482-7.
61. R. V. Jean and D. Barabé. *Symmetry in Plants*. World Scientific, 1998. ISBN: 978-981-02-2621-3.
62. H. Jönsson et al. 'An Auxin-Driven Polarized Transport Model for Phyllotaxis'. *Proceedings of the National Academy of Sciences of the United States of America* 103.5 (2006), pp. 1633–1638. https://doi.org/10.1073/pnas.0509839103.
63. S. Katok. *Fuchsian Groups*. Chicago Lectures in Mathematics. Chicago University Press, 1992. ISBN: 978-0-226-42582-5.
64. S. Katok and I. Ugarcovici. 'Symbolic Dynamics for the Modular Surface and Beyond'. *Bulletin of the American Mathematical Society* 44.1 (2007), pp. 87–132. ISSN: 0273-0979, 1088-9485. https://doi.org/10.1090/S0273-0979-06-01115-3. URL: https://www.ams.org/bull/2007-44-01/S0273-0979-06-01115-3 (visited on 28/01/2025).
65. J. Kepler. *The Six-Cornered Snowflake, 1611*. Oxford University Press, 2014. ISBN: 978-0-19-871249-7.
66. D. Kierzkowski et al. *Mechanical Interactions between Tissue Layers Underlie Plant Morphogenesis*. 3rd July 2024. https://doi.org/10.21203/rs.3.rs-4536561/v1. URL: https://www.researchsquare.com/article/rs4536561/v1 (visited on 13/03/2025). Pre-published.
67. C. Kuhlemeier. 'Phyllotaxis.' *Current Biology* 27.17 (Sept. 2017), R882–R887. ISSN: 0960-9822. https://doi.org/10.1016/j.cub.2017.05.069.pmid:28898658. URL: http://europepmc.org/abstract/med/28898658 (visited on 15/02/2019).
68. M. Kunz and F. Rothen. 'Phyllotaxis or the Properties of Spiral Lattices III. An Algebraic Model of Morphogenesis'. *Journal de Physique I* 2.11 (1992), pp. 21312172. https://doi.org/10.1051/jp1:1992273. URL: https://hal.archives-ouvertes.fr/jpa-00246692.
69. H. W. Lee and L. S. Levitov. 'Universality in Phyllotaxis: A Mechanical Theory'. In: *Symmetry in Plants*. Ed. by R. V. Jean and D. Barabé. World Scientific, 1998, pp. 619–653. ISBN: 978-981-02-2621-3.

70. B. Leisering. 'Die Verschiebungen an Helianthusköpfen Im Verlaufe Ihrer Entwickelung Vom Aufblühen Bis Zur Reife.' *Flora oder Allgemeine Botanische Zeitung* 90 (1902), pp. 378–431. URL: https://www.zobodat.at/publikation_articles.php?id=395498 (visited on 03/02/2025).
71. B. Leisering. 'Zur Frage Nach Den Verschiebungen an Helianthus-Köpfen.' *Berichte der Deutschen Botanischen Gesellschaft* 20 (1902), pp. 613–624. URL: https://www.zobodat.at/publikation_articles.php?id=386378 (visited on 03/02/2025).
72. D. L. Lentz et al. 'Sunflower (*Helianthus annuus L.*) as a pre-Columbian domesticate in Mexico'. *Proceedings of the National Academy of Sciences* 105.17 (2008), pp. 6232–6237. https://doi.org/10.1073/pnas.0711760105.
73. L. S. Levitov. 'Fibonacci Numbers in Botany and Physics: Phyllotaxis'. *Journal of Experimental and Theoretical Physics Letters* 54.9 (1991), pp. 542–545. URL: http://www.jetpletters.ac.ru/ps/1265/article_19148.shtml (visited on 15/02/2019).
74. M. Linardić and S. A. Braybrook. 'Towards an Understanding of Spiral Patterning in the Sargassum Muticum Shoot Apex'. *Scientific Reports* 7.1 (1 24th Oct. 2017), p. 13887. ISSN: 2045-2322. https://doi.org/10.1038/s41598-017-13767-5. URL: https://www.nature.com/articles/s41598-017-13767-5 (visited on 11/09/2023).
75. M. B. Linford. 'Fruit Quality Studies. II. Eye Number and Eye Weight.' *Pineapple Quarterly* 3 (1933), pp. 185–188.
76. P. P. Majumder and A. Chakravarti. 'Variation in the Number of Ray- and Disk- Florets in Four Species of Compositae'. *Fibonacci Quarterly* 14 (1976), pp. 97–100.
77. J. Marcou. *Life, Letters, and Works of Louis Agassiz*. New York: Macmillan, 1896. URL: https://www.biodiversitylibrary.org/bibliography/1810 (visited on 20/02/2025).
78. C. C. Martinez et al. 'Left-Right Leaf Asymmetry in Decussate and Distichous Phyllotactic Systems'. *Philosophical Transactions of the Royal Society B: Biological Sciences* 371.1710 (19th Dec. 2016), p. 20150412. ISSN: 0962-8436, 1471-2970. https://doi.org/10.1098/rstb.2015.0412. URL: https://royalsocietypublishing.org (visited on 30/01/2020).
79. *Mathematical Modelling in Plant Biology*/Richard J. Morris, Editor. Cham, Switzerland: Springer, 2018. ISBN: 978-3-319-99070-5. URL: https://cam-ldls.lib.cam.ac.uk/ark:/81055/vdc_100070699898.0x000001 (visited on 13/03/2025).
80. G. J. Mitchison. 'Phyllotaxis and the Fibonacci Series'. *Science* 196.4287 (1977), pp. 270–275. https://doi.org/10.1126/science.196.4287.270.
81. G. Mosca et al. 'Modeling Plant Tissue Growth and Cell Division'. In: *Mathematical Modelling in Plant Biology*. Ed. by R. J. Morris. Cham: Springer International Publishing, 2018, pp. 107–138. ISBN: 978-3-319-99070-5. URL: https://doi.org/10.1007/978-3-319-99070-5_7 (visited on 13/03/2025).
82. R. Negishi, K. Sekiguchi and K. Takahata. 'Determining Parastichy Pairs for Florets and Seeds on Sunflowers'. *Forma* 37.3 (2022), pp. 39–49. https://doi.org/10.55653/forma.2022.003.002.
83. A. C. Newell and M. F. Pennybacker. 'Fibonacci Patterns: Common or Rare?' *Procedia IUTAM* 9 (2013), pp. 86–109. https://doi.org/10.1016/j.piutam.2013.09.009.
84. T. Okabe. 'Biophysical Optimality of the Golden Angle in Phyllotaxis'. *Scientific Reports* 5 (2015), pp. 15358–15358. https://doi.org/10.1038/srep15358.
85. T. Okabe. 'The Riddle of Phyllotaxis: Exquisite Control of Divergence Angle'. *Acta Societatis Botanicorum Poloniae* 85.4 (31st Dec. 2016). ISSN: 2083-9480. https://doi.org/10.5586/asbp.3527. URL: https://pbsociety.org.pl/journals/index.php/asbp/article/view/asbp.3527 (visited on 15/02/2019).
86. J. H. Palmer. 'The Physiological Basis of Pattern Generation in the Sunflower'. In: *Symmetry In Plants*. Ed. by R. V. Jean and D. Barabé. World Scientific, 1998, pp. 145–169. ISBN: 978-981-02-2621-3.
87. M. F. Pennybacker, P. D. Shipman and A. C. Newell. 'Phyllotaxis: Some Progress, but a Story Far from Over'. *Physica D: Nonlinear Phenomena* 306 (2015), pp. 48–81. https://doi.org/10.1016/j.physd.2015.05.003.
88. H. Poincaré. *Papers on Fuchsian Functions*. Trans. by J. Stillwell. Springer, 1985. ISBN: 978-0-387-96215-3.

89. J. H. Priestley and L. Scott. 'The Vascular Anatomy of Helianthus Annuus'. *Proceedings of the Leeds Philosophical and Literary Society. Scientific section.* 3 (1936), pp. 159–73.
90. P. Prusinkiewicz and A. Lindenmayer. *The Algorithmic Beauty of Plants*. Springer, 1991. ISBN: 0-387-97297-8.
91. P. Prusinkiewicz and A. Runions. 'Computational Models of Plant Development and Form'. *New Phytologist* 193.3 (2012), pp. 549–569. https://doi.org/10.1111/j.1469-8137.2011.04009.x.
92. P. Prusinkiewicz et al. 'Modeling Plant Development with L-Systems'. In: *Mathematical Modelling in Plant Biology*. Ed. by R. J. Morris. Cham:Springer International Publishing, 2018, pp. 139–169. ISBN: 978-3-319-99070-5. https://doi.org/10.1007/978-3-319-99070-5_8. URL: https://doi.org/10.1007/978-3-319-99070-5_8 (visited on 13/03/2025).
93. P. Prusinkiewicz et al. 'Phyllotaxis without Symmetry: What Can We Learn from Flower Heads?' *Journal of Experimental Botany* 73.11 (2nd June 2022), pp. 3319–3329. ISSN: 0022-0957.https://doi.org/10.1093/jxb/erac101. URL: https://doi.org/10.1093/jxb/erac101 (visited on 26/02/2025).
94. D. Reinhardt and E. M. Gola. 'Law and Order in Plants - the Origin and Functional Relevance of Phyllotaxis'. *Trends in Plant Science* (25th May 2022). ISSN: 1360-1385. https://doi.org/10.1016/j.tplants.2022.04.005. URL: https://www.sciencedirect.com/science/article/pii/S1360138522001261 (visited on 13/07/2022).
95. D. Reinhardt et al. 'Regulation of Phyllotaxis by Polar Auxin Transport'. *Nature* 426.6964 (2003), pp. 255–260. https://doi.org/10.1038/nature02081.
96. F. J. Richards. 'Phyllotaxis: Its Quantitative Expression and Relation to Growth in the Apex'. *Philosophical Transactions of the Royal Society of London. Series B, Biological Sciences* 235.629 (1951), pp. 509–564. https://doi.org/10.1098/rstb.1951.0007.
97. F. J. Richards. 'The Geometry of Phyllotaxis and Its Origin'. *Symposium of Society for Experimental Biology* 2 (1948), pp. 217–245.
98. J. N. Ridley. 'Ideal Phyllotaxis on General Surfaces of Revolution'. *Mathematical Biosciences* 24 (December 1984 1984), pp. 1–24.
99. G. W. Ryan, J. L. Rouse and L. A. Bursill. 'Quantitative Analysis of Sunflower Seed Packing'. *Journal of Theoretical Biology* 147.3 (1990), pp. 303–328. https://doi.org/10.1016/S0022-5193(05)80490-4.
100. K. F. Schimper. *Beschreibung Des Symphytum Zeyheri*. Heidelburg: CF Winter, 1835.
101. A. A. Schneiter and J. F. Miller. 'Description of Sunflower Growth Stages'. *Crop Science 21* 21 (1981), pp. 901–903.
102. J. C. Schoute. 'On Whorled Phyllotaxis. IV. Early Binding Whorls'. *Recueil des Travaux Botaniques Néerlandais* 35 (1938), pp. 415–558.
103. J. C. Schoute. 'Uber Pseudokonchoiden'. *Recueil des Travaux Botaniques Neerlandais* 10 (1913), pp. 326–339.
104. R. Schwartz. *Mostly Surfaces*. Vol. 60. Student Mathematical Library. American Mathematical Society, 2011. URL: https://bookstore.ams.org/stml-60/ (visited on 28/01/2025).
105. H. A. Schwarz. 'Ueber Diejenigen Fälle, in Welchen Die Gaussische Hypergeometrische Reihe Eine Algebraische Function Ihres Vierten Elementes Darstellt.' *Journal für die reine und angewandte Mathematik* 75 (1873), 292-335, Taf. II. ISSN: 0075-4102; 1435-5345/e. URL: https://eudml.org/doc/148203 (visited on 21/09/2021).
106. S. Schwendener. *Gesammelte botanische Mittheilungen*. Gebrüder Borntraeger, 1898. 496 pp.
107. S. Schwendener. *Mechanische Theorie Der Blattstellungen*. Engelmann, 1878.
108. B. Shi and T. Vernoux. 'Patterning at the Shoot Apical Meristem and Phyllotaxis'. *Current Topics in Developmental Biology* (2019). https://doi.org/10.1016/bs.ctdb.2018.10.003. URL: https://www.researchgate.net/publication/329215386_Patterning_at_the_shoot_apical_meristem_and_phyllotaxis (visited on 15/02/2019).
109. B. Shi et al. 'Feedback from Lateral Organs Controls Shoot Apical Meristem Growth by Modulating Auxin Transport'. *Developmental Cell* 44.2 (22nd Jan. 2018), 204-216.e6. ISSN: 1534-5807. https://doi.org/10.1016/j.devcel.2017.12.021. URL: https://www.cell.com/developmental-cell/abstract/S1534-5807(17)31038-9 (visited on 25/09/2019).

110. R. S. Smith et al. 'A Plausible Model of Phyllotaxis'. *Proceedings of the National Academy of Sciences of the United States of America* 103.5 (2006), pp. 1301–1306. https://doi.org/10.1073/pnas.0510457103.
111. M. Snow and G. R. S. Snow. 'Experiments on Phyllotaxis II. The Effect of Displacing a Primordium'. *Philosophical Transactions of the Royal Society of London. Series B* 222 (1933), pp. 353–400. https://doi.org/10.1098/rstb.1932.0019.
112. J. Stillwell and H. Poincaré. 'Translator's Note: Poincare's Theory of Fuchsian Groups'. In: *Sources of Hyperbolic Geometry*. American Mathematical Society, 1996. ISBN: 978-1-4704-3878-4. URL: http://ebookcentral.proquest.com/lib/cam/detail.action?docID=4908553 (visited on 08/03/2022).
113. S. Strauss et al. 'Phyllotaxis: Is the Golden Angle Optimal for Light Capture?' *New Phytologist* 225.1 (2020), pp. 499–510. ISSN: 1469-8137. https://doi.org/10.1111/nph.16040. URL: https://onlinelibrary.wiley.com/doi/abs/10.1111/nph.16040 (visited on 13/03/2025).
114. J. Swinton. *Disk-Stacking Models Are Consistent with Fibonacci and Non-Fibonacci Structure in Sunflowers*. 8th July 2024. arXiv: 2407.05857[q-bio]. URL: http://arxiv.org/abs/2407.05857 (visited on 09/07/2024).
115. J. Swinton. 'Watching the Daisies Grow: Turing and Fibonacci Phyllotaxis'. In: *Alan Turing: His Work and Impact*. Ed. by C. A. Teuscher. 2nd colour. Springer, 2004, pp. 477–498.
116. J. Swinton, E. Ochu and Museum of Science and Industry Turing's Sunflowers Consortium. 'Novel Fibonacci and Non-Fibonacci Structure in the Sunflower: Results of a Citizen Science Experiment'. *Royal Society Open Science* 3.5 (2016), p. 160091. https://doi.org/10.1098/rsos.160091. URL: https://royalsocietypublishing.org/doi/full/10.1098/rsos.160091.
117. Theophrastus. *Enquiry Into Plants Book VI*. Heinemann: Loeb Classical Library, 1916. 39-39.
118. R. L. Thomas. 'Orthostichy, Parastichy and Plastochrone Ratio in a Central Theory of Phyllotaxis'. *Annals of Botany* 39 (1975), pp. 455-89.
119. D. W. Thompson. *On Growth and Form*. Cambridge University Press, 1917.
120. J. H. M. Thornley. 'Phyllotaxis. II A Description in Terms of Intersecting Logarithmic Spirals'. *Annals of Botany* 39.3 (1975), pp. 509–524. https://doi.org/10.1093/oxfordjournals.aob.a084962.
121. J. Traas. 'Phyllotaxis'. *Development* 140.2 (15th Jan. 2013), pp. 249–253. https://doi.org/10.1242/dev.074740. URL: http://dev.biologists.org/content/140/2/249 (visited on 15/02/2019).
122. S. C. Tucker. 'Phyllotaxis and Vascular Organization of the Carpels in Michelia Fuscata'. *American Journal of Botany* 48.1 (1961). https://doi.org/10.2307/2439596.
123. A. M. Turing. 'The Chemical Basis of Morphogenesis'. *Philosophical Transactions of the Royal Society of London* 237.641 (1952), pp. 37–72. https://doi.org/10.1098/rstb.1952.0012.
124. A. M. Turing. 'The Morphogen Theory of Phyllotaxis'. In: *Alan Turing: His Work and Impact*. Ed. by S. B. Cooper and J. van Leeuwen. Elsevier, 2013, pp. 773–826.
125. S. Vajda. *Fibonacci and Lucas Numbers, and the Golden Section: Theory and Applications*. Dover Publications, 2008. ISBN: 0-486-46276-5.
126. F. M. J. Van der Linden. 'Creating Phyllotaxis: The Dislodgement Model'. *Mathematical Biosciences* 100.2 (1st July 1990), pp. 161–199. ISSN: 0025-5564. https://doi.org/10.1016/0025-5564(90)90039-2. URL: https://www.sciencedirect.com/science/article/pii/0025556490900392 (visited on 03/12/2021).
127. G. van Iterson Jr. *Mathematische Und Mikroscopisch-Anatomische Studien Über Blattstellungen, Nebst Betraschung Über Den Schalebau Der Milionlinen*. Gustav-Fischer Verlag, 1907.
128. G. van Iterson Jr. 'New Studies on Phyllotaxis'. *Proceedings of the Koninklijke Nederlandse Akademie van Wetenschappen* 63 (1960), pp. 137–150.
129. C. W. Wardlaw. 'Phyllotaxis and Organogenesis in Ferns'. *Nature* 164.4161 (July 1949), pp. 167–169. https://doi.org/10.1038/164167a0. URL: https://www.nature.com/articles/164167a0 (visited on 26/09/2019).
130. A. Weisse. 'Die Zahl Der Randblüthen an Compositenköpfchen in Ihrer Beziehung Zur Blattstellung Und Ernährung'. *Jahrbücher für Wissenschaftliche Botanik*. 30 (1897), pp. 453–483.

131. A. Weisse. 'Sketch of the Mechanical Hypothesis of Leaf-Position'. In: *Organography of Plants, Especially of the Archegoniata and Spermaphyta*. Ed. by K. Goebel. Oxford: Clarendon Press, 1900, pp. 1–296. https://doi.org/10.5962/bhl.title.3802. URL: https://www.biodiversitylibrary.org/bibliography/3802

132. C. Wright. 'The Most Thorough Uniform Distribution of Points about an Axis'. *Runkle's Mathematical Monthly,* 1.VII (1859), pp. 244–248.

133. F. R. Yeatts. 'A Growth-Controlled Model of the Shape of a Sunflower Head'. *Mathematical Biosciences* 187.2 (Feb. 2004), pp. 205–221. ISSN: 0025-5564. https://doi.org/10.1016/j.mbs.2003.09.002. pmid: 14739085.

134. T. Yonekura et al. 'Mathematical Model Studies of the Comprehensive Generation of Major and Minor Phyllotactic Patterns in Plants with a Predominant Focus on Orixate Phyllotaxis'. *PLOS Computational Biology* 15.6 (6th June 2019), e1007044. https://doi.org/10.1371/journal.pcbi.1007044. URL: https://journals.plos.org/ploscompbiol/article?id=10.1371/journal.pcbi.1007044 (visited on 13/10/2019).

135. S. Yoshida et al. 'Genetic Control of Plant Development by Overriding a Geometric Division Rule'. *Developmental Cell* 29.1 (14th Apr. 2014), pp. 75–87. ISSN: 1534-5807. https://doi.org/10.1016/j.devcel.2014.02.002.pmid:24684831. URL: https://www.cell.com/developmental-cell/abstract/S1534-5807(14)00095-1 (visited on 14/10/2021).

136. B. Zagórska-Marek. 'Phyllotactic Patterns and Transitions in Abies Balsamea'. *Canadian Journal of Botany* 63 (1985), pp. 1844-54.

137. B. Zagórska-Marek and M. Szpak. 'The Significance of γ-and λ-Dislocations in Transient States of Phyllotaxis: How to Get More from Less - Sometimes!' *Acta Societatis Botanicorum Poloniae* 85.4 (31st Dec. 2016). https://doi.org/10.5586/asbp.3532. URL: https://pbsociety.org.pl/journals/index.php/asbp/article/view/asbp.3532 (visited on 26/02/2019).

138. T. Zhang, F. Wang and P. Elomaa. 'Repatterning of the Inflorescence Meristem in Gerbera Hybrida after Wounding'. *Journal of Plant Research* 134.3 (1st May 2021), pp. 431–440. ISSN: 1618-0860. https://doi.org/10.1007/s10265-021-01253-z. URL: https://doi.org/10.1007/s10265-021-01253-z (visited on 18/08/2023).

139. T. Zhang et al. 'Phyllotactic Patterning of Gerbera Flower Heads'. *Proceedings of the National Academy of Sciences* 118.13 (30th Mar. 2021), e2016304118. https://doi.org/10.1073/pnas.2016304118. URL: https://www.pnas.org/doi/full/10.1073/pnas.2016304118 (visited on 26/02/2025).

Index

C
Co-prime integers, 13
   Bézout relation, 15, 32
   winding-number pair, 16
Cylindrical lattice, 31
   classification by principal parastichy numbers, 57
   classified by principal parastichies, 46
   divergence, 31
   golden, 42
   golden angle divergence, 44
   hexagonal, 51
   multijugate, 53
   non-opposed, 46, 61
   opposed, 46
   packing efficiency, 69
   renormalisation, 65
   rhombic-tiling generalisation, 131
   rise, 31
   spiral, 47
   square, 52
   touching circle, 52

E
Euclidean algorithm, 16
   matrix form, 17
   reduction of generating pairs, 48
   van Iterson diagram, 69

P
Pairs of parastichy vectors, 36
   columnar pair, 111
   generating pair, 32, 37
   opposed pair, 37
   principal pair, 29, 32, 45
Parastichy lines, vii, 34
   as spirals, 81
   origin-parastichy, 34
   transformed, 77
Parastichy number, vii, 4, 29
   first two as lattice label, 46
   for non-lattice patterns, 128
   principal, 46
Parastichy vector, 32
   complementary vector, 32, 36
   generating vector, 32
   orthostichy vector, 50
   periodicity vector, 32
   principal vector, 46
   visible vector, 35
Placement model
   cellularly explicit, 118
   Douady-Couder, 113
   energy-based, 116
   L-system, 120
   mechanical, 119
   protractor, 119
   reaction-diffusion, 119
   stacked-disk, 123
   Turing, 117

V
Van Iterson diagram, 57
   branch, 57
   Euclidean algorithm, 69
   Fibonacci branches, 63
   triple-point bifurcation, 60, 61

© The Editor(s) (if applicable) and The Author(s), under exclusive license to Springer Nature Switzerland AG 2025
J. Swinton, *Mathematical Phyllotaxis*, Surveys and Tutorials in the Applied Mathematical Sciences 17, https://doi.org/10.1007/978-3-031-94013-2

MIX
Papier aus verantwortungsvollen Quellen
Paper from responsible sources
FSC® C105338

If you have any concerns about our products,
you can contact us on
ProductSafety@springernature.com

In case Publisher is established outside the EU,
the EU authorized representative is:
**Springer Nature Customer Service Center GmbH
Europaplatz 3, 69115 Heidelberg, Germany**

Printed by Libri Plureos GmbH
in Hamburg, Germany